PHYSICS RESEARCH AND TECHNOLOGY

FUSION ENERGY AND POWER

APPLICATIONS, TECHNOLOGIES AND CHALLENGES

PHYSICS RESEARCH AND TECHNOLOGY

Additional books in this series can be found on Nova's website
under the Series tab.

Additional e-books in this series can be found on Nova's website
under the e-book tab.

PHYSICS RESEARCH AND TECHNOLOGY

FUSION ENERGY AND POWER

APPLICATIONS, TECHNOLOGIES AND CHALLENGES

LIONEL ROMERO
EDITOR

nova publishers
New York

Library of Congress Cataloging-in-Publication Data

ISBN: 978-1-63482-548-1

Published by Nova Science Publishers, Inc. † New York

CONTENTS

PREFACE

The worldwide fusion community continues its research efforts on magnetic confinement as the most promising, long-term, environmentally-friendly power source. Despite the ongoing fusion research efforts in many countries, the technology and materials-related challenges remain formidable and will hinder and delay the first fusion demonstration plant for decades. In this book, the current understanding of technology-related challenges facing fusion research are explored. Advances in fusion neutronics integral experiments in the benchmark mock assemblies for the blanket of a fusion-fission hybrid energy reactor are also described in brief. Cold Fusion (CF) is examined as well, with the authors' argument backed by evidence that cold fusion (CF) can become more understandable, and hence more enabling to engineering, especially control engineering. The final chapter details the Force Free Helical Reactor (FFHR) and its implications on fusion power.

Chapter 1 – Advances in fusion neutronics integral experiments in the benchmark mock assemblies for the blanket of a fusion-fission hybrid energy reactor is described in brief. The concept of the hybrid energy reactor is that fusion energy from magnetic confinement fusion device is multiplied by released fission energy from the subcritical reactor. A conceptual blanket with the subcritical reactor is mainly loaded with natural uranium and lithium material. In the fission fuel region, uranium material and light water are arranged alternately. The integral experiments are performed for validating nuclear data and code in conceptual design. Based on materials available, the mock assemblies of the blanket are established. The spherical fission assemblies for fission research and plutonium production consist of three layers of depleted uranium shells and several layers of polyethylene shells. The spherical lithium assembly for tritium production consists of depleted uranium and LiPb alloy shells. The cubic fission assembly consists of natural uranium and polyethylene and its structure is basically consistent with one of the fission fuel region. In the assemblies with the D-T fusion neutron source, the plutonium production rates, uranium fission rates, ratios of capture to fission, ^{238}U (n, 2n) reaction rates and tritium production rates are measured, separately. The measured results are compared to the calculated ones with MCNP code and ENDF/B-VI library data.

Chapter 2 – The worldwide fusion community continues its research efforts on magnetic confinement as the most promising, long-term, environmentally-friendly power source. A wealth of experimental evidence has been amassed with the goal of controlling and sustaining the fusion process. However, the technology advances have been only related to the confinement physics experiments, but not towards developing the relevant power plant technologies. This chapter addresses our current understanding of technology-related challenges facing fusion research: liquid metal and ceramic breeder blankets, sufficient tritium breeding in blankets to fuel the plasma, radiation-resistant structural materials for 14-MeV neutron environment, appropriate thermo-hydraulic design especially in the first wall to allow efficient electrical energy production, advanced manufacturing techniques to fabricate complex components, maintenance schemes relating to plant design and availability, and viable approaches to handle the mildly radioactive fusion materials. Despite the ongoing fusion research efforts in many countries, the technology and materials-related challenges remain formidable and will likely hinder and delay the first fusion demonstration plant for decades. Strong, well-funded technology-based R&D programs must be established to solve the critical issues.

Chapter 3 – This chapter argues that Cold Fusion (CF) can become more understandable, and hence more amenable to Engineering, especially Control Engineering, if we will take account of appropriate redevelopments in Physics. These redevelopments become possible with fuller application of standard Mathematics. Of particular interest is the venerable protocol of formulating governing differential equations, identifying an appropriate family of solutions for the differential equations, and applying the boundary conditions that specify the particular problem at hand. This protocol leads to review of some results from Special Relativity Theory (SRT), which leads in turn to a better entrée into Quantum Mechanics (QM), and hence into Chemistry and Electro-chemistry, and finally into CF. The authors find the following practical recommendations re CF: 1) Let the heavy water rest quietly for a long time before attempting to elicit fusion events; 2) Consider both gravity and electrochemistry to anticipate how various ion species will distribute themselves in the water; 3) Study a range of possible arrangements of electrodes in the water to determine the optimum one.

Chapter 4 – Super high dense core (SDC) plasma with more than $1x10^{21}$ m^{-3} observed in the large helical device (LHD) stimulates the creation of concept of controlling the unstable operating point in the high density and low temperature regime for the Force Free Helical Reactor (FFHR). This novel control method discovered in 2006 can stabilize the unstable operating point using proportional- integral-derivative (PID) control of the fueling with the error of the fusion power of $e(P_f)= -(P_{fo}-P_f)$. Here $P_{fo}(t)$ is the fusion power set value and $P_f(t)$ is the measured fusion power. Although continuous fueling with the small density variation has been actively studied at the unstable operating point, it is still uncertain whether the operating point can be stabilized when discrete fueling such as pellet injections is used. In this work, the authors show at first that it is possible to control the unstable ignited operation for the density variations corresponding to the pellet size of 12 mm~16 mm in FFHR2m (R=14 m, $\bar{a}$ =1.73 m). In addition, in the commercial reactor FFHR2m2 (R=15.7 m, $\bar{a}$ =2.5 m,) the external heating power is specified to be the foreseeable neutral beam injection (NBI) system with the energy of 1.5 MeV. In FFHR2m2 the initial density is kept at the low value less than ~$4x10^{20}$ m^{-3} to overpass the saddle point on Plasma Operation Contour (POPCON) map by the neutral beam injection heating. After passing over the saddle point, a high-density ignition is accessed by alpha heating after turning off NBI injection. Here the control algorithm of the NBI external heating power has been newly implemented based on the temperature limit concept instead of the preprogramming used in FFHR2m. Various disturbances in plasma parameters due to impurity injection and confinement factor change can also be stabilized to some extent at the thermally unstable point by the feedback control of the heating power and discrete fueling.

In: Fusion Energy and Power
Editor: Lionel Romero

ISBN: 978-1-63482-548-1

Chapter 1

TECHNOLOGY-RELATED CHALLENGES FACING FUSION POWER PLANTS

Laila A. El-Guebaly[1,*]**, *Lorenzo V. Boccaccini***[2]**,**
Richard J. Kurtz[3] ***and Lester M. Waganer***[4]

[1]University of Wisconsin, Fusion Technology Institute, Madison, WI, US
[2]Karlsruhe Institute of Technology, Karlsruhe, Germany
[3]Pacific Northwest National Laboratory, Richland, WA, US
[4]O'Fallon, MO, US

ABSTRACT

The worldwide fusion community continues its research efforts on magnetic confinement as the most promising, long-term, environmentally-friendly power source. A wealth of experimental evidence has been amassed with the goal of controlling and sustaining the fusion process. However, the technology advances have been only related to the confinement physics experiments, but not towards developing the relevant power plant technologies. This chapter addresses our current understanding of technology-related challenges facing fusion research: liquid metal and ceramic breeder blankets, sufficient tritium breeding in blankets to fuel the plasma, radiation-resistant structural materials for 14-MeV neutron environment, appropriate thermo-hydraulic design especially in the first wall to allow efficient electrical energy production, advanced manufacturing techniques to fabricate complex components,

* Corresponding author: E-mail: elguebaly@engr.wisc.edu.

maintenance schemes relating to plant design and availability, and viable approaches to handle the mildly radioactive fusion materials. Despite the ongoing fusion research efforts in many countries, the technology and materials-related challenges remain formidable and will likely hinder and delay the first fusion demonstration plant for decades. Strong, well-funded technology-based R&D programs must be established to solve the critical issues.

1. Introduction

Beginning in the early 1950's and continuing to the present, scientists have examined a multitude of fusion concepts searching for the ultimate source of energy with a viable confinement approach. The magnetically confined tokamak is currently regarded as the most feasible concept for energy generation as its program accounts for over 90% of the worldwide magnetic fusion effort. The current wealth of tokamak information and understanding includes more than 40 operational experiments, approximately 30 worldwide tokamak power plant and DEMO studies, extensive physics R&D programs with impressive international collaboration in all areas of fusion research, and a large body of accumulated knowledge [1]. However, these achievements are insufficient to conclusively demonstrate the capability of producing commercial fusion power. This is due to the minor effort devoted to the nuclear technology development relative to burning plasma science. This fusion research imbalance does not hinder the current experiments, but creates a major impediment for fusion power plants and will certainly delay the construction of the first fusion DEMO for decades. The fusion technology and materials-related challenges are generally known and remain unresolved due to the low priority assigned to their developments and the lack of 14 MeV neutron testing facility.

This chapter characterizes selected technology-related challenges that needs to be solved to achieve commercially viable fusion energy: efficient liquid metal and ceramic breeder blankets, sufficient tritium breeding in blankets to fuel the plasma, low activation/radiation-resistant structural materials for 14 MeV neutron environment, appropriate thermo-hydraulic capabilities in the first wall and blanket to allow efficient electrical energy production, advanced manufacturing techniques to fabricate complex and reliable components, effective maintenance schemes relating to plant design and availability, and viable and sustainable approaches to handle the mildly radioactive fusion materials during operation and after decommissioning.

Other technologies covered in companion chapters and in literature are deemed essential for the development of integrated fusion power plants include the divertor, vacuum vessel, shielding components, magnets, heating and current drive equipment, heat transfer and transport system, tritium and deuterium fuel handling and storage, instrumentation and control, etc.

In a typical tokamak design (shown in Figure 1), the first wall, divertor, and heating/current drive (H/CD) launchers are plasma-facing components subject to 14 MeV neutrons, charged particles, and radiation from plasma, presenting the most challenging fusion components. The materials in these components will be activated, eroded, and damaged by highly energetic neutrons and charged particles, requiring remote maintenance and frequent replacement during plant operation. The following paragraphs summarize the functionality of selected tokamak components along with our current understanding of technology-related challenges.

First Wall and Blanket Materials and Design – Behind the first wall is the blanket that breeds the tritium (a fuel element for sustained plasma operation), attenuates the neutrons and gamma rays, and captures and converts their kinetic energies into heat used to produce electricity. The tritium breeding and heat removal are demanding tasks that require careful selection of the breeder materials (liquid metal or ceramic) and coolants (water, liquid metal or helium) and the arrangement of such materials within the blanket (Section 2).

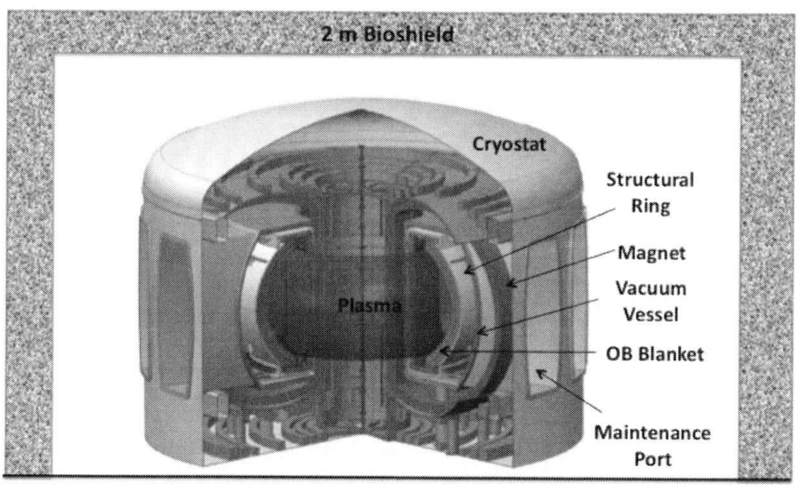

Figure 1. Typical tokamak design depicting essential fusion power core components, cryostat, and biological shield.

Tritium Self-Sufficiency – Any fusion device will have to generate its own tritium (T) fuel as T does not exist in nature in any appreciable quantity. The tritium breeding ratio (TBR) represents a metric for T self-sufficiency. The required TBR (> 1) is design and breeder dependent and should be established and carefully tailored for each blanket design separately. The breeding margin (TBR – 1) accounts for known deficiency in nuclear data, approximation in 3-D blanket modeling, and T holdups, losses, and decay (Section 3).

Structural Materials – Structural materials derived from the fission industry are inadequate for fusion due to the more damaging effects of the 14 MeV neutrons. Radiation-resistant low-activation steels need to be specifically developed, tested and qualified for fusion applications. The currently favored structural material is a ferritic/martensitic steel that is suitable for moderate operating temperatures up to 550°C. There is a need to develop superior and more capable structural materials [such as the nano-scale oxide dispersion strengthened alloys that could offer higher temperature strength, microstructural stability, He-trapping capability, and radiation damage resistance. Specific radiation damage limits are not yet based on any 14 MeV exposure data, but these data could be achieved with directed R&D programs coupled with a 14 MeV neutron facility (Section 4).

Advanced Manufacturing – After the power core has been designed, there remains a substantial set of engineering challenges to manufacture the complex geometries with new classes of materials while achieving a low cost part that is highly reliable and meets all the nuclear code regulations. New and innovative fabricating and joining technologies offer solutions, such as additive manufacturing of structural components, highly automated processes, direct deposition of multiple materials, and tailoring of the structure to accommodate local loads and stresses. Demonstration, pilot, and/or prototype plants will affirm the maturity and applicability of these technologies for future fusion power plants (Section 5).

Robotic Maintenance – The degradation of materials limits the lifetime of the fusion power core elements thus requiring periodic and unscheduled maintenance of components. The levels of radiation within the power core are sufficiently elevated to not allow any human exposure inside the bioshield barrier. Thus all operations inside the bioshield must be accomplished by remote means. The power core and the maintenance approach are intimately integrated to produce an approach that will achieve high power plant availability (85-95% is the long term goal). This goal demands a very efficient maintenance procedure. The replacement of larger sections offers the promise of higher reliability and maintainability capabilities. Horizontal removal of large sectors of the power core allows the quick replacement of failed sectors. High accuracy and precision placement can be accomplished with autonomous, computer-guided robotic equipment (Section 6).

Handling Fusion Radioactive Materials – There is some latitude in the selection of low-activation materials for fusion designs to prevent creating any high-level waste (HLW), but the amount of low-level waste (LLW) generated will be quite large. Developing and demonstrating processes and procedures to recycle (i.e., reuse within the nuclear industry) and clear (i.e., release to the commercial market if containing traces of radioactivity) all materials are essential to enable fusion deployment, reclaiming of resources (through less mining of ore), and maintaining a sustainable environment (Section 7).

2. First Wall and Blanket Challenges

The first wall (FW) is the largest plasma-facing component. It covers 85-90% of the power core inner surface and is only 5-15 cm away from the 100 million °C plasma surface. It is subject to harsh fusion environment consisting of: energetic 14 MeV neutrons, intense radiation from the plasma with a steady surface heat load, H/He blistering, sputtering, and surface erosion by high velocity charged ions (D, T, and He). These extreme conditions damage the structural materials and degrade their mechanical and physical properties. Infrequent plasma disruptions with large electro-magnetic forces, and the more frequent, but less intense, off-normal events of high energy discharges, such as edge localized modes and vertical displacement events, can also damage the FW elements. The few centimeter-thick, bare FW structure is not sufficiently robust to withstand the high power loading. A thin (2-5 mm) tungsten (W) layer is presently considered as a favorable armor to protect the FW. However the highly energetic neutrons penetrate beyond the FW and in the blanket subsystem where they will displace atoms, transmute the original materials and generate He and H gases within these materials. Such damaging conditions degrade the FW/blanket structure during operation and create a limited lifetime. This requires periodic maintenance to replace the FW/blanket every 5-10 years in future power plants, depending on the neutron wall loading (see Section 6). More near-term, less radiation-resistant materials used in large experimental facilities may experience shorter lifetimes.

The highly energetic neutrons can penetrate beyond the FW/blanket, thus activating all materials in the fusion power core. Low-activation materials are carefully selected for all fusion components to minimize the generation of high-level waste. Currently, significant efforts are underway to develop more radiation-resistant structural materials that could operate in a higher temperature regime relative to the first generation of reduced-activation ferritic/martensitic (RAF/M) steels that limit the FW/blanket operating temperature to less than 550°C (refer to Section 4). Currently, F82H, EUROFER, and CLAM are the favored RAF/M steels for steel-based fusion blankets, offering an operational temperature window between 350°C and 550°C. The embrittlement of steel by irradiation limits the lower end of the operating window, while the thermal creep limits the upper end.

Other attractive FW/blanket structural materials include vanadium alloys and SiC_f/SiC composites. Breeding materials can be either solid or liquid materials containing lithium or lithium compounds. The likely coolant media choices include water, helium, liquid metal or molten salt systems. In addition to the variation in material choices, the FW and blanket design approaches will impact all aspects of the fusion power core and will require extensive integration between all components and subsystems. In particular, the FW heat removal aspect represents a serious challenge because of the unpredictable geometric peaking of the heat loads and the need to efficiently integrate the FW cooling with remaining power generation systems. In most fusion designs, the FW is an integral element of the blanket. Its coolant must be compatible with the structural material and exhibit low chemical energy releases when interacting with blanket materials (breeder and/or neutron multiplier) during an accident. To enhance the overall thermal conversion efficiency (η_{th}), the FW/blanket (and divertor) must operate at high temperatures (> 500°C).

In any fusion design, the breeding of tritium is absolutely necessary, but it represents a major challenge to the blanket design (refer to Section 3). Several Li-based breeding systems were conceived and developed to some degree over the past 40-50 years to help breed tritium: liquid breeders including Li, LiPb, Flibe (LiF-BeF_2), Flinabe (LiF-NaF-BeF_2), and LiSn, and ceramic (or solid) breeders including Li_2O, Li_4SiO_4, Li_2ZrO_3, Li_2TiO_3, and $LiAlO_2$. Due to their low breeding capability, all ceramic breeders require the addition of beryllium or lead neutron multiplier to provide the required level of tritium breeding for sustained plasma operation (net TBR >1.0). The lithium used in almost all liquid metal and ceramic breeders needs to be enriched in ^6Li to 20-90% – above the natural content of 7.5 atom%. Currently, the cost of Li enrichment is unknown and there is no commercial process to supply large quantities of enriched Li (e.g., ~40 tons of enriched Li is needed for about 7000 tons of PbLi in ARIES-ACT-1).

Liquid metals (Li, PbLi, and LiSn) and molten salts (Flibe and Flinabe) can serve as a tritium breeder and/or coolant. As will be discussed shortly, some liquid metals (LM) suffer from magneto-hydrodynamic effects in the presence of a strong magnetic field, limiting their heat transfer capability as a coolant. Such LMs are supplemented with water or He coolant as in water-cooled lead lithium (WCLL), He-cooled lead lithium (HCLL), and dual-cooled lead lithium (DCLL) blanket concepts.

Many RAF/M steel-based ceramic and LM blankets employ the pressurized helium coolant (8-10 MPa) to achieve sufficiently high heat transfer capability with tolerable pumping power, He exit temperature of ~500°C, and $\eta_{th} \leq 40\%$. Thermal conversion efficiency up to 45 % is only feasible with the DCLL blanket where the PbLi exit temperature reaches ~650°C. More advanced blankets employing SiC_f/SiC composites or advanced steels could operate at higher temperatures (> 700°C), offering higher $\eta_{th} >$ 45%. The operating parameters for the water coolant is limited to ~325°C, 15.5 MPa, and η_{th} of 33%, based on experience from existing fission industry. However, supercritical water (with higher temperatures and pressures) could achieve higher η_{th} (~45%).

In ceramic breeder blankets, water is the main coolant for water-cooled ceramic breeder (WCCB) blankets while He is the coolant of choice for He-cooled pebble bed (HCPB) blankets. Pumping a He purge gas through the ceramic breeders is an essential element of all ceramic breeder blankets. The tritium formed in the ceramic pebbles has to migrate to the surface to be picked up by the He purge gas. In the case of LM/MS blankets, no migration of T is required as the flowing breeder circulates outside the fusion power core, transporting the T produced within the breeder; then the T is extracted from a side stream of the LM/MS coolants. In DCLL blankets, the T control is facilitated by the high circulation rate of PbLi. This limits the T partial pressure to less than 1 Pa (<< 100 Pa in HCLL and WCLL blankets). Tritium from any breeder can permeate through the structure and coolant piping, and eventually accumulate in the He and water coolants of WCCB, HCPB, WCLL, HCLL, and DCLL blankets, which is undesirable. For safety reasons, the T inventory in the water and He coolants should be as-low-as-reasonably-achievable. Therefore, continuous online detritiation is mandatory for both coolants. Because this process is very expensive (taking into account the amount of coolant involved), the T permeation into coolants has to be limited to only few percent of the total T production. An efficient detritiation system is a strong requirement, not only for coolants, but also for all facility areas exposed to tritium.

In summary, the FW/blanket subsystem faces several technological challenges: material behavior under irradiation, compatibility between materials, T breeding capability, T behavior, control and extraction, efficient heat removal scheme, negative impact of strong magnetic fields on LM flow, Li enrichment process, efficient detritiation system, fabrication of complex geometries, periodic replacement and maintenance, and safety and economic constraints. The following sections address in more detail some of these challenges.

2.1. Basic Functions and Design Requirements

The FW and blanket provide two prime functions: breed all the tritium needed for plasma fueling and recover the neutrons and radiation energies at high temperatures. Additionally, the FW and blanket provide a shielding function by slowing down and attenuating the majority of neutrons, thus reducing the neutron exposure to the external components: vacuum vessel, shield, and magnets. To enhance the thermal conversion efficiency, the FW and blanket must operate at high temperature and thus employ high temperature coolants to recover the neutron and radiation energies at a reasonable coolant pressure level. A technical concern is the ability of the coolant in the FW and blanket cooling channels to efficiently remove the surface and volumetric heating without requiring significant steel structure (that compromises the breeding) and pumping power (that degrades the power balance).

Since the neutron-induced damage to the blanket structure decreases with the radial distance from the plasma core, the sizable outboard (OB) blanket could be segmented into two distinct zones: an inner life-limited segment to be replaced several time during operation, and an outer lifetime segment that requires no (or fewer) replacements. Such a segmentation help reduce the blanket radwaste stream and replacement cost, but may add complexity to the design and its maintenance operation.

An additional requirement for the blanket is to accommodate the Kink and Vertical Stabilizing shells, as advanced plasma physics mandates embedding such shells in the OB blanket. The accommodation, support, and cooling of such shells present a technical challenge. Hence, segmenting the blanket offers another advantage – the ability to accommodate these shells in the space between OB blanket segments. The key features and design requirements for advanced FW/blanket designs suitable for power plants are summarized in the following points:

- Innovative FW design to handle > 2 MW/m^2 neutron wall loading
- Tritium breeding capability (overall TBR > 1.05)
- Online adjustment of breeding for LM/MS breeders
- High temperature operating regime (> $500^{\circ}C$) to achieve high thermal conversion efficiency and enhance the power balance
- Radiation-resistant, low-activation structural materials to survive > 150 displacement per atom (dpa) and achieve attractive FW/blanket lifetimes
- Compatible blanket constituents with low chemical energy releases during accidents
- Mechanically and chemically stable ceramic breeders with low T retention, wide operational temperature window, and simple fabrication and recycling processes
- Advanced fabrication techniques to improve economics and reliability
- Fabrication method and behavior of thermal/electrical insulators for LM blankets
- Minimal magneto-hydrodynamic effect on liquid breeder flow, pumping power, and heat transfer
- Recyclable breeder and structural materials to minimize waste assigned to geologic disposal
- Blanket segmented into replaceable and permanent components to reduce the radwaste stream and replacement cost
- Blanket to accommodate W shells needed for plasma stabilization
- Efficient tritium extraction systems
- Efficient tritium containment, control, and processing
- Efficient detritiation system for water and He coolants
- Highly reliable components to minimize unscheduled outages
- Maintainable components with easy access to shorten downtime.

2.2. Liquid Breeder Blankets

Breeding blankets have been designed with a variety of liquid metals and molten salts, such as natural lithium, lead lithium, Flibe, Flinabe, or lithium tin. The lead and beryllium serve as a neutron multiplier to enhance the tritium breeding of PbLi and Flibe/Flinabe breeders, respectively. The lithium in PbLi, Flibe, Flinabe, and LiSn compounds is normally enriched in ^{6}Li as necessary to achieve the required TBR for a particular blanket design application. To avoid excessive corrosion rates, the maximum interface temperature between LM and RAF/M steel must remain between 470°C and 520°C, depending on the LM flow velocity and piping configuration. Each of these LM/MS blanket concepts offers advantages and disadvantages with a common goal of satisfying the basic requirements mentioned above.

From the outset of the fusion program in the early 1970s, pro-fusion countries (initially the U.S., Europe, Russia, and Japan; joined more recently by China, S. Korea, and India) have explored all blanket options and developed blanket designs characterized by breeding materials, structural materials, and coolants. Some blanket designs are more advanced than others, suggesting various methods for blanket installation and replacement. Currently, each country focuses on developing the preferred blanket concept and establishing the required R&D program.

In principle, liquid breeders have the following features:

- Pro - Tritium self-sufficiency without additional neutron multiplier, except Flibe/Flinabe
- Pro - Online adjustment of lithium isotopes can maintain required TBR
- Pro - High operating temperature leading to high thermal conversion efficiency
- Pro - Immunity to radiation damage
- Pro - Relatively low operating pressure
- Pro - High thermal conductivity compared to ceramic breeders
- Pro - Tritium extraction from the breeder performed outside the blanket in low radiation environment
- Con - Corrosive to steel structure
- Con - LMs are magnetically conductive in strong magnetic fields and require electric insulator.

The more widely studied LM/MS breeder materials are Li, PbLi and Flibe /Flinabe. Lithium is compatible with vanadium alloys [2, 3] that offer higher maximum temperature of the structure (700°C) compared to RAF/M steels (550°C). PbLi is compatible with two structural materials: 1) SiC_f/SiC composites where the PbLi exits the blanket at very high temperature

(~1100°C) [4-6], and 2) RAF/M structure where the PbLi exits the blanket at a lower temperature (~ 700°C) [6, 7].

Three main approaches are envisioned for LM/MS blankets:

1. A self-cooled blanket where Li [2, 3], Flibe/Flinabe [8, 9], and PbLi [4-6] circulate at a high flow speed to remove the nuclear heating
2. A quasi-stagnant LiPb circulates slowly and exits the blanket for T extraction while the majority of the heat is removed with water or He coolants, as in the WCLL blanket [5, 10-13] and HCLL blanket [5, 14-17]
3. A dual-cooled PbLi (DCLL) blanket [5-7, 17-19] where the FW and blanket structure are cooled with high pressure He (8 MPa). The breeding zone is cooled by circulating a low-speed PbLi to external heat exchanger and tritium extraction systems. The flowing PbLi removes about 50 % of the total heat and is electrically and thermally insulted from the load carrying walls by flow channel inserts preferably made from SiC_f/SiC composites.

In the U.S., the DCLL blanket is the preferred option for future devices: FNSF [20, 21], DEMO, and power plants [6, 7]. Internationally, the Europeans are currently examining the HCLL, DCLL, and WCLL blankets. Russia is pursuing the Li/V/Be blanket [3], Japan prefers the Flibe/Flinabe breeders [9], while China is developing both HCLL and DCLL blankets [17]. The following subsections cover the technology challenges facing selected LM/MS blanket concepts: Flibe/Flinabe, WCLL, HCLL, and DCLL.

2.2.1. Self-Cooled MS Blankets

Flibe is the breeder of choice for the self-cooled MS blanket of the FFHR design [9]. Flibe was also examined during the APEX project [8]. Salient features of the Flibe blanket include the low magneto-hydrodynamic pressure drop, low reactivity with water and air, low operating pressure, and low hydrogen solubility that helps reduce the T inventory in the blanket. Any Flibe blanket design cannot achieve TBR > 1 [22] without dissolving additional beryllium multiplier in Flibe (as in the FFHR blanket design [9]) or adding a thin layer (4 cm) of lead multiplier (as in the APEX blanket design [8] shown in Figure 2). There is a severe corrosion issue with the release of free fluorine after tritium is bred, as fluorine forms a tritiated hydrofloric acid, which is highly corrosive to all steel-based structural materials. Beryllium has been considered as a REDOX agent for chemistry control. A design concern is the narrow temperature window, limited by the Flibe high melting point (~360°C) and maximum temperature of ferritic steel (550°C). To reduce the T permeation through the heat exchanger, mixing hydrogen-absorption metal powders (e.g., Ti, Zr, or V) with Flibe was proposed [9]. To widen the operating temperature window to 305-700°C, the FFHR study suggested the use of low-activation vanadium alloy (that offers higher operating temperature (700°C) than steel) along with the Flinabe breeder (that has lower melting point (305°C) than Flibe). The thermal conversion efficiency could reach 40% for the Flinabe/V blanket option [9] and 46% in the APEX design that employs the advanced nano-structured ferritic steel [8].

2.2.2. Water and Helium-Cooled PbLi Blankets

The use of water to cool PbLi blankets [10-13] requires similar coolant conditions as in fission pressurized water reactors (15.5 MPa pressure, 325°C maximum temperature with ~33% η_{th}). Since RAF/M steels become brittle below 350°C, the WCLL blanket concept could only be viable if new RAF/M variant is developed for a reduced susceptibility to radiation hardening and ductile fracture behavior in the 250-350°C regime during irradiation with fusion neutrons [23]. The high potential for a chemical reaction between water and PbLi raises safety concerns: hydrogen generation, temperature increase, over-pressurization of the blanket box, and the possibility of hydrogen ingress into the vacuum vessel if the blanket box fails. Figure 3 displays a full OB module for the original 1995 WCLL blanket design where the water coolant flows in tubes to remove the heat from the PbLi breeder [11, 12]. Alternatively, smaller modules with straight surfaces could be attached to a common back supporting structure housing the feeding pipes [13]. Double wall tubes are necessary to reduce the probability of water leakage within the module and minimize the probability of water/PbLi interaction.

The European HCLL blanket utilizes the RAF/M EUROFER structure, PbLi breeder, and He coolant with inlet/outlet temperature of 300/500°C and 8 MPa pressure [14-16]. The limitations of the envisaged blanket concepts are mainly due to materials. The temperature window of the EUROFER structure limits the He temperature and thus directly impacts the efficiency of the power conversion system (< 40%). This efficiency could improve with the development of oxide dispersion strengthened (ODS) ferritic steels that allow higher operating temperatures, exceeding 750°C. Figure 4 illustrates a typical HCLL blanket design [15].

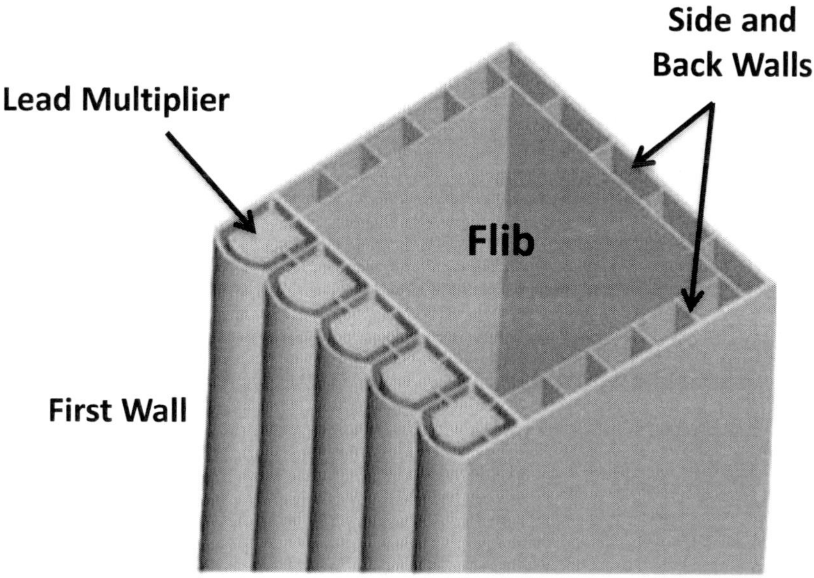

Figure 2. Cross section of APEX Flibe blanket design [8].

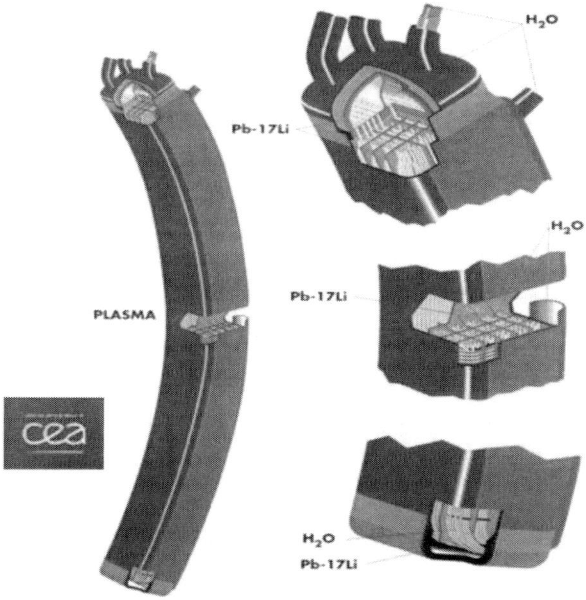

Figure 3. Outboard module of WCLL blanket [11, 12].

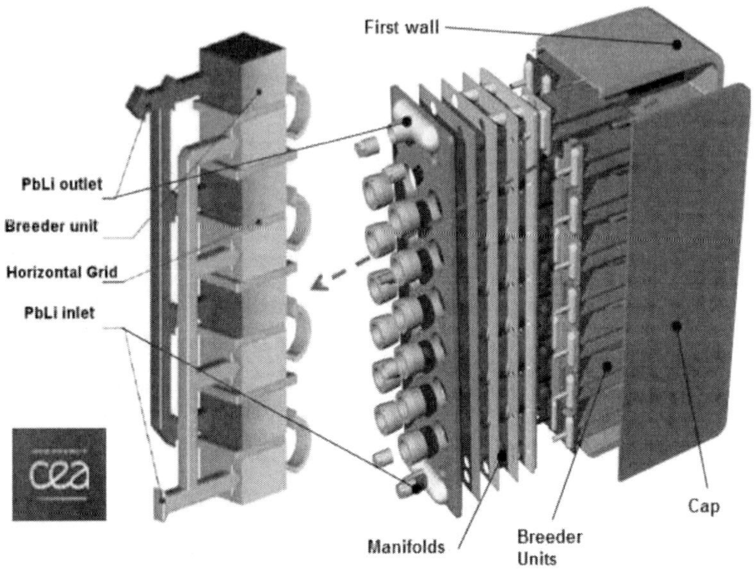

Figure 4. HCLL blanket design for ITER testing module [15].

A common concern for all PbLi-based blankets (WCLL, HCLL, and DCLL) is the low solubility of T in PbLi that could lead to considerably large T partial pressure and consequently high T permeation flux into the cooling water or helium. Since the allowable T concentration level in circulating coolants must be low for safety reasons, a continuous online extraction of T is necessary for both WCLL and HCLL blankets. Such a detritiation process is also required, to a less extent, for the DCLL blanket, but the extraction of T from water is much more difficult, very expensive, and consumes considerable amount of electricity. Note that WCCB blankets (Subsection 2.3) share the same problems facing the WCLL blankets, regarding the embrittlement of RAF/M steel below 350°C and safety concerns, as steam reacts vigorously with the beryllium multiplier in case of an accident.

2.2.3. Dual-Cooled PbLi Blankets

The DCLL blanket is the most popular LM blanket concept worldwide. A unique feature of the DCLL blanket [5-7, 17-19] pertains to the exit PbLi breeder/coolant temperature that exceeds the RAF/M steel maximum temperature (allowed by a clever design approach). This innovative idea was originally proposed by Malang [24] in the 1990s for the U.S. ARIES spherical tokamak (ST) design [7] and applied ever since to subsequent DCLL blanket designs developed in the U.S. [6, 19-21] and in other countries [5, 17, 18]. The main idea suggests decoupling the hot PbLi breeder/coolant from the RAF/M steel using a flow channel insert (FCI). Basically, the FCI thermally insulates the RAF/M steel structure from the hotter PbLi and electrically insulates the conductive PbLi from the RAF/M steel structure. At present, the SiC FCI is the prime candidate for the DCLL blanket. Back-up options include steel-encase alumina, spinel, or zirconia that offer limited temperature windows. An isometric view of a typical DCLL blanket design with SiC FCI is shown in Figure 5.

The PbLi LM is conductive and, in the presence of tokamak and ST strong magnetic fields, appreciable magneto-hydrodynamic (MHD) forces will occur within the moving PbLi. In the absence of an electrical insulation between the flowing LM and the load carrying duct walls, such MHD forces will drag and slow the PbLi flow, inducing unfavorable flow velocity profiles and resulting in large pressure drops and high mechanical stresses in duct walls. The thermal flow channel inserts [25, 26] serve also as an electrical insulator to alleviate the MHD pressure drop in the blanket, inlet/outlet pipes, manifolds, and ducts. It is the most critical element of the DCLL blanket design. A blanket-related R&D program is needed to characterize the MHD pressure drop and performance of the 0.5 cm thick SiC thermal/electrical insulator in fusion relevant environment.

The more recent ARIES-ACT-2 power plant design [27, 28] describes the details of the DCLL blanket and outlines three main challenges facing this design: the MHD effect on flow and heat transfer, the tritium containment and control, and the fabrication and behavior of the RAF/M steel and FCI under irradiation. In this design, the blanket breeds sufficient T with a wide margin (refer to Section 3), the PbLi exits the blanket at ~650°C, the maximum RAF/M steel temperature is below 550°C, while the maximum steel/PbLi interface temperature remains below 500°C to avoid excessive PbLi/steel corrosion. The 650°C PbLi exit temperature allows the use of the Brayton cycle with thermal conversion efficiency of 45%.

In the future, more advanced steel (such as the high Cr nano-structured ODS steel) could be alloyed with Al (and Zr) to improve the PbLi/ODS corrosion resistance for applications at 700-800°C [29]. The PbLi could exit the blanket at 750°C and η_{th} could potentially approach 50%.

Continuous extraction of T from a side stream of the flowing PbLi helps keep the circulating T inventory low (< 50 grams). A promising T extraction method is the use of a permeator device [30] where the PbLi flows through tubes located in a vacuum chamber. The T diffuses from the PbLi layer through a tube that has high T permeability (such as Nb or Ta), and is then extracted by a vacuum pump. An R&D program would involve demonstrating such a T extraction method, determining the efficiency of the T permeator design, and addressing other critical issues related to T losses into the He coolant and environment.

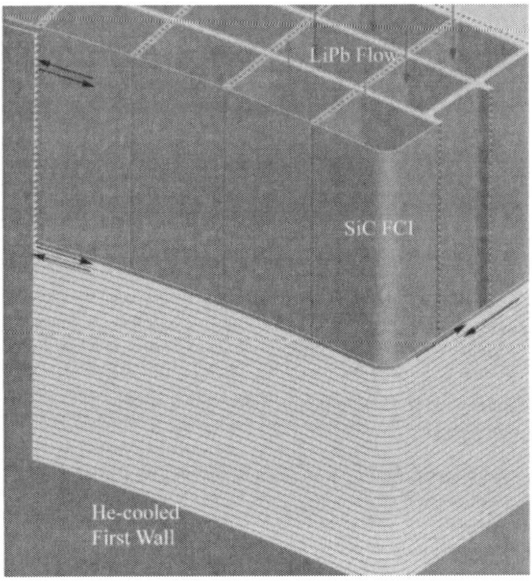

Figure 5. Typical DCLL blanket isometric showing the FW (dark grey), radial/toroidal cooling channels (light grey), breeding zones surrounded with flow channel inserts (orange).

2.3. Ceramic Breeder Blankets

Solid breeder blanket concepts with beryllium are an attractive option for breeding blanket offering high potential of tritium breeding due to the exceptionally high neutron multiplication factor of beryllium. Drawbacks are the high damage rate of breeder and beryllium under irradiation and strong thermal fluxes. Both functional materials show a rapid deterioration of mechanical properties under neutron irradiation (e.g., embrittlement and burn-up) and break easily under thermal gradients or swelling. For this reason the most studied concepts make use of these materials in the form of pebble beds with pebble diameters lower than 1 mm. In this form, these materials are structurally sound with relatively stable properties (in particular the thermal conductivity) during the entire irradiation time. Furthermore, the small dimension of pebbles limits pebble fragmentation and tolerance. In other words, a large number of pebbles could crack without impacting the purge or thermal properties of beds. However, the use of pebble beds challenges the thermal design of the breeder zone, as the resulting thermal conductivity is low (only few W m^{-1} K^{-1} for ceramic bed) [31].

Compared to liquid breeding blankets, pebbles beds in solid breeding blankets are designed to remain in the blanket for the entire blanket lifetime. This means they have to withstand the accumulated thermal and irradiation damage, as they cannot be replaced during operation. Furthermore, over the lifetime of the blanket, the lithium content in the ceramic breeder will be depleted due to Li burn-up thus reducing the neutronic performances. Also the blanket structural materials will be damaged by irradiation and thermal stresses (e.g., creep, fatigue, etc.). All these adverse phenomena limit the lifetime of the component to few years of irradiation (depending on the neutron flux). The DEMO could operate under 70 dpa neutron damage to blanket steel. The expected characteristics of the breeder material seem to be satisfactory and the solid breeder burn-up deterioration can be accommodated in the initial neutronic design. It is desirable to increase the lifetime performance, but the ultimate limits of almost all materials are still unknown

The list of solid materials considered for the tritium breeding is not large; it includes Li-oxide for the high density of Li that can favor more T breeding and compact design, but also ternary Li-ceramics that present higher mechanical stability under a large thermal operational window; typically proposed ternary ceramics are ortho-silicate (Li_4SiO_4), meta-titanate (Li_2TiO_3), meta-zirconate (Li_2ZrO_3) and aluminate (γ-$LiAlO_2$). Requirements that can guide in the selection are, in addition to Li density and mechanical stability, low tritium retention, chemical and micro-structural stability in wide operational conditions, simple fabrication and recycling processes, low activation under irradiation, as well as chemical compatibility with structural materials. None of the mentioned materials can satisfy the whole list of requirements, however a preference has grown for Li_4SiO_4 and Li_2TiO_3 that are at the present the reference choice of almost all the proposed solid breeder concepts. The current worldwide effort is to produce and characterize these two ceramic compounds [32].

Beryllium is an essential component of a solid breeder blanket; it provides the required neutron multiplication and can be used in solid form as pebble at high temperature. However, its use in breeding blanket creates several issues. Maximum temperature has to be limited (usually lower than 650°C) as mechanical properties deteriorate at high temperatures causing swelling problem. Also, if the rate of T production in Be during neutron irradiation is negligible for the total T breeding, T can be trapped in He bubbles that form during the same process. This presents a safety issue if the T amount exceeds the T inventory limits for plant operation. However, recent evaluations of in-pile experiment have shown that for highly irradiated materials the release rate of T should be sufficiently high to avoid safety concerns in power plants [33]. Be intermetallic compounds have been also investigated to improve the Be performance in terms of T release, mechanical properties and mitigation of steam reaction. An alternative compound (used in Japanese designs) is $Be_{12}Ti$ ("beryllide") [34].

The present concepts of solid breeder blankets were formed with large conceptual studies in the 1990s, in particular in Europe and Japan. In Europe, Dalle Donne proposed a Breeder-Out-of-Tube (BOT) concept for NET [35]; a helium cooled concept based on pebble beds. Some years later, this concept became a reference concept in the EU Breeding Blanket program with the name of Helium Cooled Pebble Bed (HCPB) concept for DEMONET and for the ITER TBM Program [36]. In Japan during the same time, the reference version for the TBM program became a water-cooled concept with pebble bed [37]. The present versions of the EU and JA ITER TBM and DEMO concepts are derived from these parallel developments. In the late 2000s, China and Korea joined the ITER TBM Program and developed solid breeder concepts [38, 39] with similar architecture.

The main features of these concepts are the employment of solid breeder and beryllium in the form of a flat pebble bed, the use of RAF/M steels as structural material and the use of separated circuits for the heat and tritium removal; the coolant loop (helium or water) is connected externally to the electrical energy production system while other dedicated loops (known also as "purge loops") carry the tritium produced in the beds to out-of-vessel tritium extraction systems (TES) of the outer fuel cycle of the plant with gas carrier. This division of function ensures an effective and economical extraction of T in a loop in which T can be concentrated and can be separated from the coolant loop. It improves the safety by preventing large quantities of tritium from reaching loop components in contact with the secondary power systems, e.g., steam generators. Due to the functional and physical separation of heat and tritium removal circuits, a possible T permeation from the purge to the coolant loops in the blanket box is considered a parasitic effect and has to be minimized in the design. Figure 6 illustrates the two different loops: the coolant (in this example helium at 300-500°C) loop is connected to the steam generators of a Rankine power generation system, and a coolant purification system (CPS) operates in by-pass to extract T contaminating the coolant.

Both breeder and coolant are contained in boxes that provide the mechanical support and the insulation from the vacuum chamber as well as the division of the two functional circuits. Hence, the pebbles are not in direct contact with the coolant, but the heat produced in the flat bed is removed by surrounding plates with internal channels or tubes that contain the high pressure coolant. The thickness of beds is strongly limited to only few centimeters due to the low thermal conductivity of the bed itself and the temperature limit of materials. Since the plates only cool the pebble bed externally, the maximum temperature of the pebbles is reached in the middle of the beds with strong temperature gradients. The coolant (helium or water) flows in the channels and box structures and is collected in manifold systems usually placed in the rear of the blanket behind the breeding zone to not compromise the tritium breeding capability.

These concepts make use of RAF/M steels that are under development in several countries for near term fusion application [40]. As present RAF/M productions can operate at a maximum temperature of ~550°C, and could be extended to 650°C for more advanced steels. Structural materials are limiting the maximum operating temperature that the coolant can reach in these concepts and consequently the efficiency of the electrical generation system.

As far as the coolant is concerned both helium and water have been proposed for these concepts. Water in liquid phase is an exceptional coolant offering high heat transfer performances (due to its high specific thermal capacity), requiring low pumping power; it is widely available and its cooling technology is well developed. For electricity production, the typical pressurized water reactor (PWR) conditions have been proposed for fusion power plants (285-325°C at 15.5 MPa) with thermal efficiency of ~33%. Concepts with supercritical or sub-critical water at higher pressure are also considered for advanced power plants to increase coolant temperatures and thermal performance. The issues of water in solid breeder blankets are related mainly to its compatibility with solid breeders and reaction with beryllium in case of an accidental contact at high temperature. In particular, Be reacts exothermically with steam at temperature greater than ~600°C, producing H that raises safety concerns. To the contrary, helium is compatibility with both solid breeder and beryllium; the disadvantages of helium relate to its performances as a coolant (lower heat transfer capability, low density that results in high pumping power, less mature technology, etc).

The tritium is generated inside the pebbles and has to migrate to their surface before being removed by the gas carrier. The time of residency of tritium in the pebbles (called Tritium Residence Time, TRT) is very important parameters for the design because it can dictate large part of the T inventory in the plant. The mobility of T in pebble depends on the temperature (with exponential law), geometry, type of solid breeder material, and chemistry of the purging atmosphere. For typical ortho-silicate pebbles at 500°C, the TRT is about few hours; hence essential goal of the design should be to maximize the average temperature of the solid beds in order to minimize the total T inventory in the pebbles. The extraction of T from the ceramic pebbles surface is accomplished with a purge gas at low pressure. The chemistry of the He purge gas has also an important role in the T extraction process; e.g., the addition of H favors the extraction of T per isotopic exchange, but introduces the hydrogen that must be separated later from the tritium fuel.

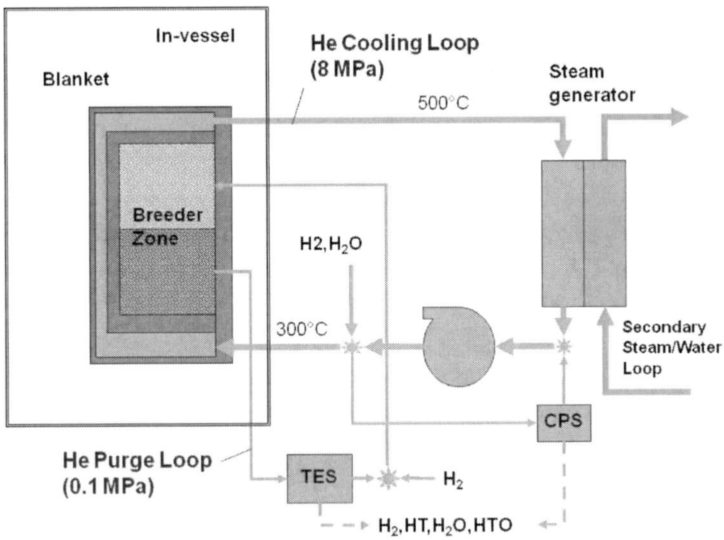

Figure 6. Conceptual scheme for the HCPB blanket.

Inside the blanket, T can escape from the purge circuit permeating into the adjacent coolant loop; as the two circuits share a very large interface surfaces (in order to adequately cool the pebble beds) and are separated by thin steel walls that operate at high temperature (reaching a maximum of 550-650°C). The amount of T that can potentially permeate through the steel wall is large. The contamination of the coolant of the primary system represents a safety issue; T can reach the external environment through the secondary loop and the T inventory can continue to build up in the coolant if not efficiently extracted. For this reason, detritiation systems for the coolant are required (e.g., CPS location for helium loops in Figure 6).

Because the water and helium coolants require completely different T extraction technologies, both versions of the solid breeder concept can largely differ in layout, allowable limits and performances. For both coolants, the necessity to design an economically viable system mandates reducing the T permeation rate under few % of the T production. It is possible to adjust the parameters of the purge flow (e.g., mass flow, pressure, chemical composition) to reduce the T permeation rate; if such adjustments are not sufficient, it is necessary to consider permeation barriers (e.g., alumina coating) at the interface. Due to many uncertainties in the surface transport parameters for T (that make the calculation unreliable) along with the lack of clear requirements and limits, the necessity of permeation barriers is still an open issue. The possibility to avoid coating layers and rely only on chemical control of the permeation surfaces would be a clear advantage for solid breeder concepts [41].

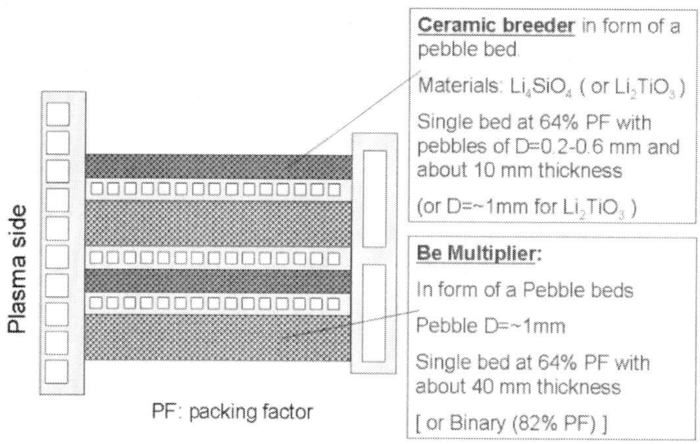

Figure 7. Conceptual scheme for the European HCPB blanket.

Figure 7 illustrates a He-cooled concept for the HCPB blanket for the 2003 design proposed for the European Power Plant Conceptual Study [42]. A He-cooled first wall facing the plasma removes the direct surface heating; inside the box, a "sandwich-type composed of cooling plates that support and cool the breeder material and Be neutron multiplier – both in a form of pebble bed. This box provides the dual function of T-breeding and removal of the volumetric nuclear heating. The beds are oriented in perpendicular direction to the FW; this configuration has been selected for the HCPB concept to increase the mechanical strength of the box (as coolant plates serve also as stiffening plates) and to facilitate the arrangement of manifold at the rear of the blanket. Both breeder and Be pebble beds are "purged" by an independent low pressure/low velocity gas flow; He (at 0.11 MPa and velocity of about 2 cm/s) has been selected as a T carrier with an addition of 0.1% H. The reference breeder material is Li_4SiO_4 with ^{6}Li enrichment of about 50%. It is use in the form of a poly-disperse bed with ~64% packing factor and a diameter between 0.2-0.6 mm. The max design temperature does not exceed 920°C in order to have enough margin to the melting point. Be is used in the form of pebble bed with 1 mm pebbles diameter. The structural material is EUROFER – a RAF/M steel version developed in the European program; the temperature window of EUROFER constrains the temperature of the He coolant to 300-500°C.

More recently, the European HCPB blanket has been developed further for application in the DEMO-2050 design [43]. Figure 8 shows the most recent design of the box and breeding zone. The robust box structure is divided in cells (about 20cmx20cmx40cm each) by horizontal and vertical stiffening grids where the breeding zone within the breeder unit is located. The breeder units is a He cooled U shape structure that contains ceramics between the cooling plates surrounded by Be beds.

The water-cooled blanket concept developed in Japan is illustrated in Figure 9. This blanket version was developed for a conceptual design of a compact fusion DEMO plant that assumes technologies foreseeable in 2020–2030 [44]. The main box architecture and the tritium extraction are similar to the helium concept, but only the water coolant replaces the helium coolant in the structure with channels and tubes. In addition, the flat pebble beds are organized in parallel layers to the FW; this design allows an optimum use of ceramic and beryllium from a neutronic and thermal point of view. Li is used in the form of orthosilicate for single beds and together with $Be_{12}Ti$ to form mixed pebbles beds; Be is used for additional single beds in front of the blanket to enhance the neutronic balance. In this study, water is used in sub-critical conditions (290-360°C, at 23 MPa); it flows in a series of tubes organized in parallel layers like the beds. The reference structural material is F82H – a RAF/M steel developed in Japan for fusion applications.

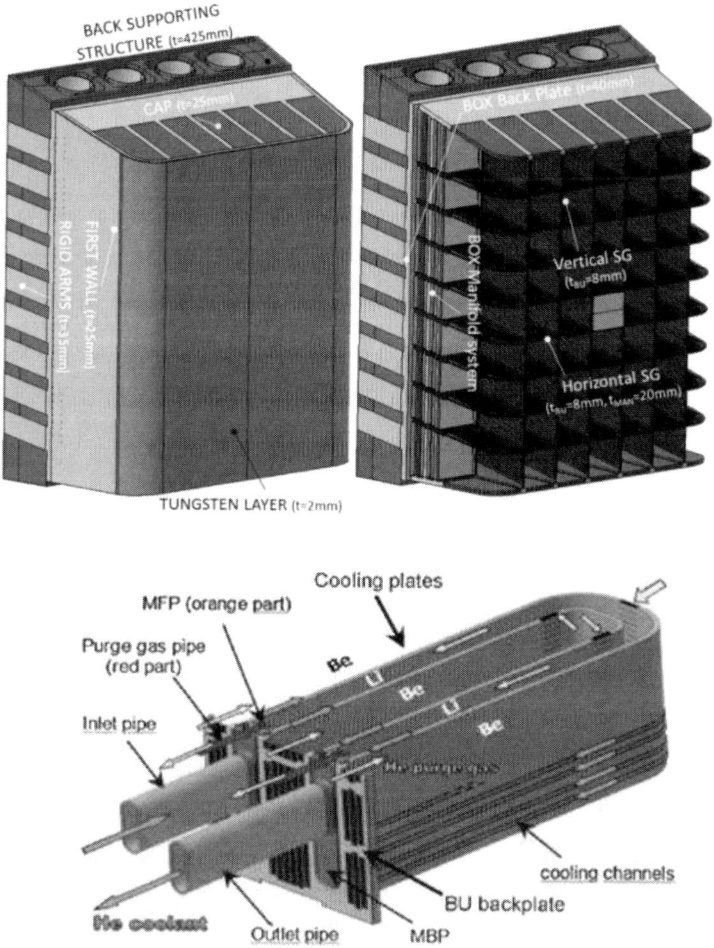

Figure 8. European HCPB blanket design for 2050 DEMO: box structure (top) and breeder unit (bottom).

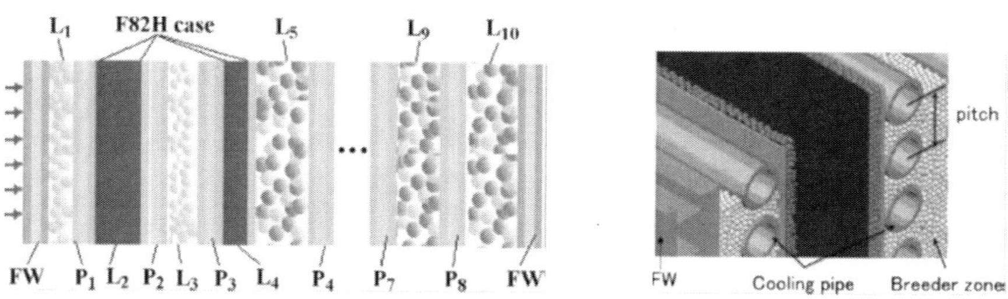

Figure 9. Breeder zone architecture for the Japanese water cooled concept in [44].

The blankets described so far are typical blankets foreseen for DEMO and first generation of fusion power plants assuming advanced technology developments, but potentially achievable during the next 25 years. Studies have been done also to investigate more advanced solutions toward the direction of higher temperatures to enhance the efficiency of the thermal-hydraulic cycle. The key to achieve this target is the development of structural materials that can be employed at high temperature, like ODS steel or very high temperature like SiC_f/SiC. An example of this potential concept was presented in 2000 [45] where the replacement of EUROFER with SiC_f/SiC and the slight increase of Be temperature limit (700°C) allow increasing the maximum He temperature to 700°C; at this temperature a He Brayton secondary cycle is allowed with gross thermal efficiency up 46%. Another example is the U.S. solid breeder blanket developed in the ARIES-CS study (see Figure 10) where the blanket makes use of the advanced ODS steel with an upper limit of 700°C, allowing helium outlet temperature of ~610°C and a Brayton cycle for the energy production with gross thermal efficiency of 43% [46].

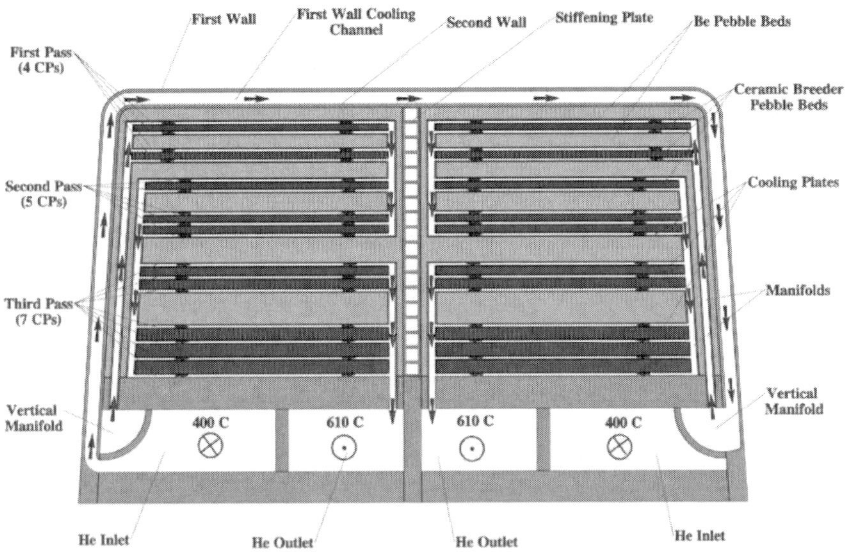

Figure 10. Bed orientation and cooling scheme of the U.S. ARIES-CS solid breeder blanket in [46].

Another scheme to increase the He coolant temperature further has been investigated using a completely different concept for heat and tritium removal.

Assuming a max temperature for Be and ceramic pebble of ~900°C, a single He circuit could fulfill the cooling and T removal functions by directly flowing He in the pebble beds [47]. The He flow scheme and overall blanket design are both simplified, but the real challenge is the thermal-hydraulic layout and materials of the unique coolant/purge loop to handle the 900°C helium flow carrying the entire T inventory for extraction. This seems unreachable to our present knowledge.

3. Challenges of Tritium Self-Sufficiency

There is a definitive need to breed all the tritium (T) required for fusion power plant operation. The tritium breeding ratio (TBR) is a key metric for tritium self-sufficiency. A net TBR above unity implies the facility breeds tritium at a level exceeding the combination of consumption, uncertainties in the calculated TBR, and T holdups, losses, decay, etc. The pertinent questions are:

- Do external sources for kilograms of T exist in the U.S. or abroad? Does a shortage of T present an economic burden or place the plant operation at risk? (Subsection 3.1)
- How high the TBR should be? (Subsection 3.2)
- How to control the overproduction of T in case of over-breeding blanket? (Subsection 3.3)
- Could R&D activities related to T breeding reduce the uncertainties in TBR prediction? (Subsection 3.4)

The following subsections will briefly answer these key questions.

3.1. Tritium Resources and Economics

The cost of purchasing T is expensive enough to enable the evaluation process in terms of defining the mission (e.g., T self-sufficiency) and designing some components to achieve the TBR-related goals. There is a worldwide interest in building several D-T fueled devices between 2020 and 2040 as plasma burning devices, technology/material testing facilities, or demonstration plants. Such fusion devices will have to either generate their own T or compete for the very limited external sources of T. Tritium does not exist in nature in any appreciable quantity. Fortunately, it can be produced from Li by capturing neutrons in fusion blankets. It is believed that ITER [48] will consume almost all T recovered from CANDU reactors (~1.7 kg/y) [49]. Tritium could be produced in small quantities using accelerators, but this option may not be economical. Other sources of T exist in the U.S. and other countries, but they are limited in supply, classified, and/or inaccessible for general use. For these reasons of limited or no T-supply, fusion devices generating thousands MW of fusion power and using hundreds kg of T annually must breed their own T to negate the risk of relying on external supplies to provide/control the essential fuel of the machine.

To provide a meaningful baseline, the TBR should be estimated with high fidelity. A typical fusion power plant with 2000 MW fusion power consumes 111 kg of T per full power year (FPY) of operation. A small 1% error in the TBR estimation is equivalent to ~1.1 kg of T/FPY. At a T unit cost ranging from ~$30k to ~$118k per gram [Ref. 50], this 1% deficiency implies an additional FPY operational cost of $33-131M to purchase T from external sources. Even larger deficiencies in TBR represent a significant burden on the operational cost, as shown in Figure 11.

3.2. Tritium Breeding Requirement

The TBR is design and breeder dependent [51]. For the dual-cooled LiPb (DCLL) blanket, a calculated TBR of 1.05 should achieve T self-sufficiency [52]. This is a computational target that could be raised or lowered, if needed, during operation with online adjustment of ^{6}Li enrichment of the PbLi breeder/coolant. The 0.05 breeding margin provides for currently known deficiencies in the nuclear data, approximations in the 3-D modeling of DCLL blanket, startup T supply for new power plant, and T holdups, losses, and decay. In the future, these unknowns should be reduced.

The current 0.05 breeding margin is relatively low for the following reasons:

- Recent integral experiments at ENEA in Italy [53] predicted < 3% uncertainties in the nuclear data of two European blanket designs: He-cooled PbLi and He-cooled pebble bed. No data available for the DCLL blanket concept
- Recent advances in University of Wisconsin computational techniques [54] enabled computing the TBR with high fidelity for the complex

configuration developed by blanket designers. This computational tool (that couples the CAD with 3-D neutronics code) enables good modeling of experimental facilities (first point above) to help obtain good agreement between calculated T production and measured values

- Excess T to provide the startup inventory for future power plant with T burnup fraction > 10%
- Aggressive assumptions were made related to minimal T inventory/holdups in all subsystems, highly efficient T extraction system with redundant components, short times for T reprocessing, efficient detritiation systems, negligible T losses to the environment, and an efficient T accountancy system.

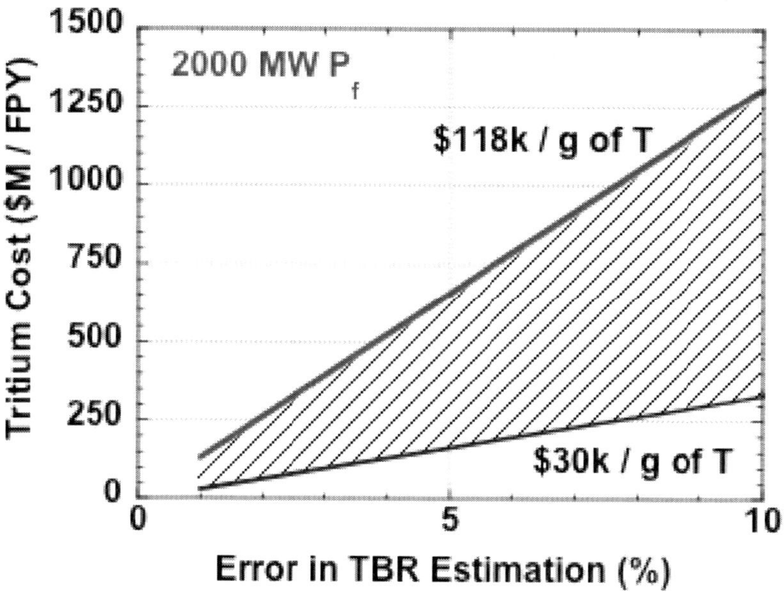

Figure 11. FPY cost of purchasing tritium from external sources versus deficiency in TBR of fusion power plant with 2000 MW fusion power.

Figure 12 displays the breeding capacity of candidate ceramic and liquid breeders. Due to low breeding capability, all ceramic breeders require beryllium metal (or Be compound) to generate the required T for sustained plasma operation. In most blanket designs, Li needs to be enriched in ^{6}Li above 10 atom%. Several design elements could significantly affect the local TBR reported in Figure 12: the FW thickness, design of side/top/bottom/back walls, choice of coolant, cooling channel geometry, flow channel insert (FCI) material and thickness, W stabilizing shells, assembly gaps, and size and orientation of H/CD penetrations. It is essential to design all in-vessel components around the TBR requirement by maximizing the blanket coverage, reducing the structural content within the FW and blanket, minimizing the H/CD aperture size and maximizing its tangential radius.

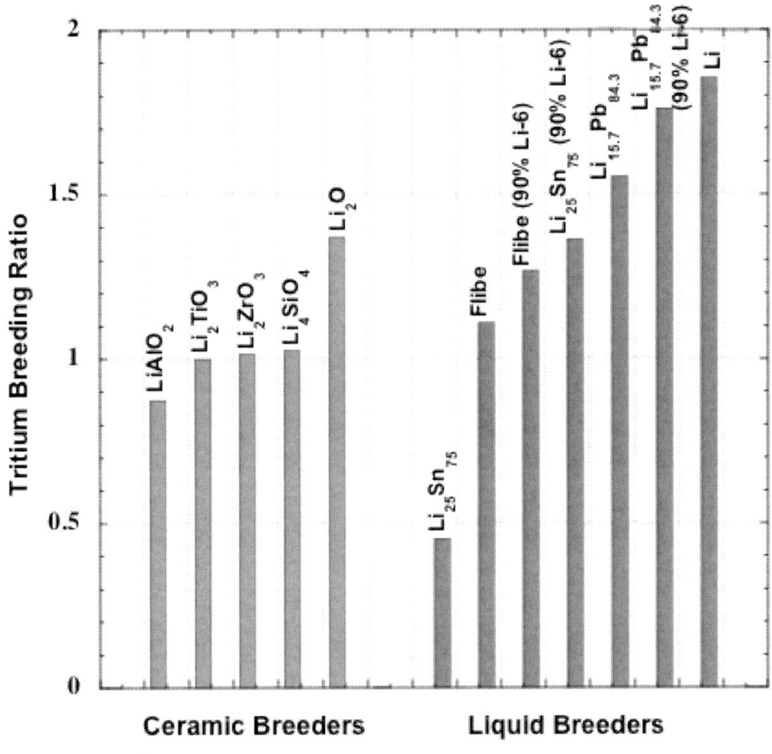

Figure 12. Local TBR for selected liquid and ceramic breeders (fairly thick breeder, no structure, no multiplier, natural Li unless indicated, 100% dense breeder at room temperature. Some breeders contain a neutron multiplier; e.g., Pb in LiPb and Be in Flibe) [51].

Past findings provide useful insight into the understanding of how the individual design elements influence the TBR and what conditions or changes affect the breeding. As an example, Figure 13 displays the stepwise degradation in the TBR of ARIES-ACT-2 [28] by various elements of the DCLL blanket internals and surroundings. The bar chart started with the theoretical upper value for TBR using only the LiPb breeding material, then each step incrementally reduces the TBR. The findings can be summarized in the following points:

- Limiting the blanket coverage radially and vertically has the largest single impact on TBR (~10%)
- The inclusion of all structural walls, cooling channels, and FCIs dropped the TBR by ~18%.
- Stabilizing shell, penetrations, and assembly gaps dropped the TBR by ~6%.
- The outboard and inboard blankets generate most of the T (69% and 28%, respectively), while the top/bottom blankets/manifolds behind the divertor provide the balance (3%).
- The T production peaks around the midplane and decreases vertically away from the midplane and radially away from the FW. This emphasizes the importance of the midplane for breeding and suggests keeping the outboard midplane free of penetrations to maximize the TBR.
- Figure 14 suggests an enrichment of 40% for ARIES-ACT-2 that results in ~15% reduction in TBR.

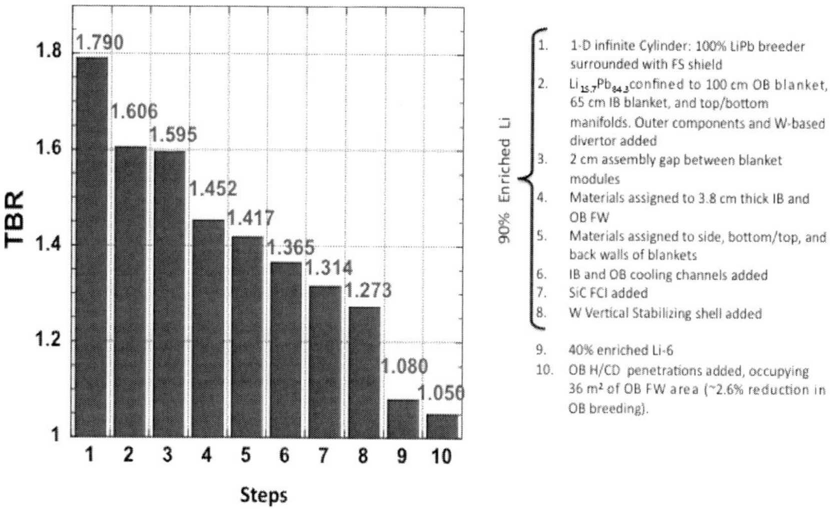

Figure 13. Bar chart showing the reduction in TBR from 1.8 to 1.05 as a result of including internals and externals of ARIES-ACT-2 DCLL blanket [28].

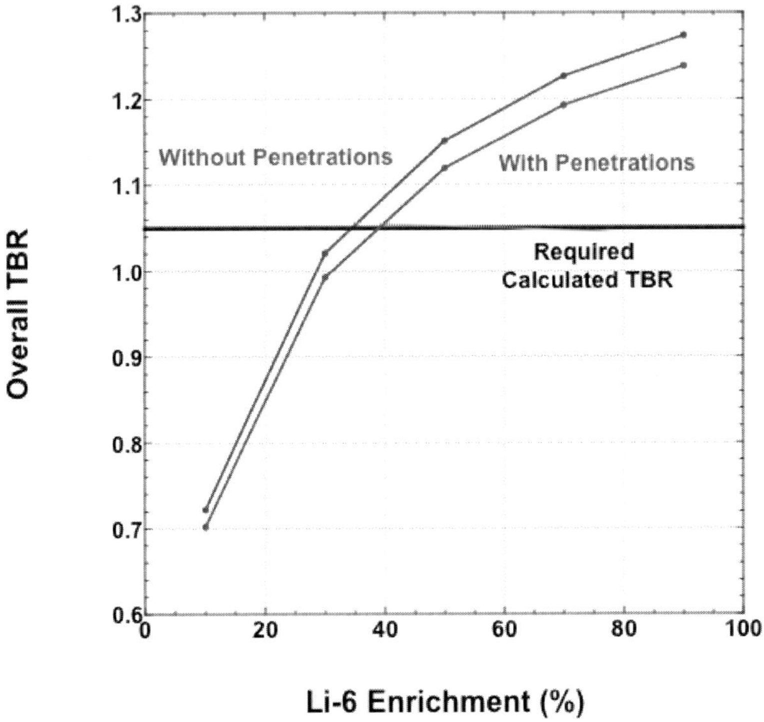

Figure 14. Tritium production distribution within LiPb breeder and change in TBR with varying ^{6}Li enrichment of LiPb breeder [28].

3.3. Need for Online Adjustment of Tritium Bred in Blanket

Despite numerous theoretical models developed to predict the dynamic of T inventory and even after the power plant is designed and built, there will be uncertainties during the facility's operation that will govern the actual breeding level. The net TBR cannot be verified until after operating the D-T fusion facility with fully integrated blanket, T extraction system, and T processing system. Thus, the definitive answer to the question of "Will the blanket overbreed or underbreed?" cannot be made before the operation of a fully integrated DEMO plant. A shortage of T would impact the plant operational schedule and mandate purchasing T from outside sources with uncertain pricing (refer to Figure 11), while a surplus of T introduces licensing and storage problems. It is also possible that fusion plants may not operate in a uniform manner, generating more/less T during operation according to the availability of the extracted T. Equally important for licensing considerations, fusion designs should not generate more T than needed for plasma fuelling. Therefore, any blanket design should have a flexible approach and be able to accept a few necessary changes in order to deliver a net TBR slightly above one.

The most practical solution is to operate with a nominal ^{6}Li enrichment < 90% and develop a scheme that adjusts the Li enrichment online during operation to compensate for unanticipated tritium production, usage and losses [55, 56]. Such a strategy is feasible for fusion designs employing liquid breeders, but is difficult to envision an approach for ceramic breeders. It is less risky to design an overbreeding blanket with a feasible scheme to lower the enrichment online shortly after plant operation. In the case of an underbreeding blanket with a net TBR <1.01, the simplest change is to increase the Li enrichment. If the increase in ^{6}Li enrichment proves to be insufficient, major design changes could involve installing blanket capability behind the divertor, operating in a single null configuration, and/or adding a beryllium multiplier. Clearly, these changes would be onerous and probably would have to be done in the design phases and not as a retrofit change.

3.4. Breeding-related R&D Program

A well-planned and executed R&D program is needed to reduce the unknowns involving the T production, storage, processing, etc. A large knowledge gap exists between near-term fusion experiments (such as ITER [48] that generates ~4 g of T/y) and a future power plant (that could produce and consume ~110 kg of T/FPY). Several small-scale T breeding-related experiments are currently operational worldwide and provide limited information. For instance, the 14 MeV neutron integral experiments (such as the Frascati Neutron Generator at ENEA Frascati in Italy [57], the Fusion Neutron Source at JAEA in Japan [58], the TUD experiment in Germany [59]) help validate the predictive capabilities (nuclear data and codes) used to calculate the TBR. Thus far, activities have been limited to the He-cooled LiPb, He-cooled pebble bed, and water-cooled ceramic breeder. Testing prototype breeding blankets in the ITER Test Blanket Modules (TBM) [60] will provide initial screening data for blanket concepts that have the potential for achieving T self-sufficiency. However, such data cannot scale up to DEMO or power plants as the limited fluence, short pulse length, small testing space, and low T production rate in the ITER TBM will not be adequate to confirm if T self-sufficiency can be achieved in DEMO with a given blanket concept. Prior to operating a DEMO, a dedicated R&D program, involving both analytical studies and lab-based demonstrations, is essential to reduce the breeding margin to ~1%, validate the computed T production rates, demonstrate the T recovery and storage processes, and determine the T inventory, holdups, and efficiency of T processing and detritiation systems.

4. Development of Damage-Resistant Structural Materials

This section discusses the challenges facing the development of robust structural materials that are able to effectively tolerate the harsh conditions of the fusion nuclear environment and meet challenging performance requirements while simultaneously satisfying the safety and environmental goals. A brief overview of the neutron and thermal loads imposed on first-wall/blanket and plasma facing materials surrounding the fusion chamber is presented. It is necessary to develop low or reduced activation structural materials to achieve an attractive fusion energy system with positive environmental, safety and economic attributes. The characteristics and critical development issues that must be resolved for leading candidate structural materials for plasma facing and first-wall/blanket applications are described. The section concludes with a discussion of the materials research needed to successfully design, construct and operate all advanced plasma devices beyond ITER.

4.1. Overview of Neutron and Thermal Wall Loadings and Radiation Damage

Compared to structural materials in commercial light-water reactors (LWR), the structural materials in proposed next-step high-energy plasma devices and demonstration fusion power plants will see much higher temperatures and experience greater neutron-induced damage. A typical LWR reactor pressure vessel operates at a temperature of about 300°C, and core internal structures experience a maximum displacement damage of less than ~50 displacements per atom (dpa) after 40 years of service [61]. In contrast, depending on the fusion confinement concept, the upper operating temperature limits for fusion blanket components will be hundreds of degrees higher than conventional fission [62].

The plasma facing structure of the first wall is only a few millimeters thick thus volumetric heating of these materials is small in comparison to surface heating. The magnitude of the surface heat flux depends on design specifics, but steady-state heat loads on important elements of a fusion power system will range from 1 to 20 MW/m^2. Such heat loads are comparable to those tolerated by rocket nozzles and will require a highly efficient heat transfer system to accommodate these enormous heat fluxes. Plasma facing and structural materials in a fusion system are also exposed to off-normal plasma events that deposit enormous quantities of heat in small regions over very time short period.

A peak neutron wall loading of ~2 MW/m^2 for the ARIES-ACT-2 power plant produces about 21 dpa per full power year (FPY) for steel components and approximately 7 dpa/FPY in tungsten [28]. In addition to causing atomic displacements, the 14 MeV neutrons from the DT reaction also interact with the atoms of structural materials to cause nuclear transmutations, or conversion of one isotope or chemical element into another. Both solid elements and gases such as helium and hydrogen are produced by these transmutation reactions. Substantially more helium and hydrogen are produced in fusion materials than in similar irradiations performed in fission reactors due to the much softer neutron energy spectrum in a fission reactor [63]. For example, for the same fast neutron fluence of 10^{27} n/m^2, the helium production in steel is only about 22 appm in a typical fission reactor compared to 810 appm in a magnetically confined fusion system [63]. Notably, the displacement damage levels are similar at ~71 dpa in a fission reactor versus ~83 dpa in a fusion device [63], but the He-to-dpa ratio is quite different: 0.3 in fission versus 10 in fusion.

The differences in operating environments between fission and fusion have significant implications for the irradiation-induced microstructural evolutionary path that structural materials will follow. Generally the much higher operating temperatures, displacement damage levels, and production of solid and gaseous transmutation products of fusion power systems relative to LWRs will cause accelerated material degradation and reduced component lifetimes. This also introduces additional variables that can confound analysis and interpretation of experiments performed in non-fusion relevant irradiation facilities.

4.2. Need for Low Activation Materials

A key challenge is the development of high-performance materials that provide an economically attractive fusion power system while simultaneously achieving the needed safety and environmental acceptability goals. Radioactive isotope inventory, and release paths are important considerations to achieving the desired safety goals. Development of low or reduced activation materials is critical to ensure that structural materials removed from service will not require long-term geological disposal and may offer the potential for recycling, thus further minimizing impact on the environment.

4.3. Leading Candidate First-Wall/Blanket and Plasma Facing Structural Materials

4.3.1. Reduced Activation Ferritic/Martensitic Steels

Reduced-activation ferritic/martensitic (RAF/M) steels are the leading reference structural materials for future fusion power plants in virtually all of the international fusion materials research programs. This is because these steels are much more technologically mature than the other candidate materials systems. The typical chemical composition of these steels is Fe-(8-9)Cr-2W-0.25V-0.07Ta-0.45Mn-0.1C, where the values are in weight percent [64]. The microstructure of RAF/M steels can be manipulated by heat treatment. Generally RAF/M steels are normalized at a temperature of about 1100°C where the material is completely austenitic [64]. The steel is transformed to martensite by quenching from the normalization temperature. A tempering heat treatment at about 760°C is then required to relieve internal stresses associated with quenching and improve mechanical properties [64]. The ability to manipulate the microstructure through heat treatment offers significant advantages to control mechanical properties. On the other hand, when heat treatments are required to restore properties following various fabrication operations this may be a detriment, especially for very large and geometrically complex components like the vacuum vessel.

The operating temperature window for RAF/M steels is approximately 200°C [65]. The lower operating temperature limit is about 325°C [66], which is controlled by irradiation-induced hardening and embrittlement phenomena. Neutron irradiation of RAF/M steels at temperatures $\leq$ 325°C can dramatically increase the yield and ultimate tensile strength, while reducing the tensile ductility. In addition, the transition temperature for ductile-to-brittle fracture can shift from an unirradiated value of around -100°C to well above room temperature [67]. Transmutation produced helium may significantly exacerbate low-temperature hardening and embrittlement as the data presented in Figure 15 suggest. The data in Figure 15 show that neutron-only displacement damage, while substantial at low irradiation temperatures, saturates at relatively modest fluence. Helium in conjunction with displacement damage can cause additional hardening and embrittlement as the data in Figure 15 shows, but these data were obtained under non-prototypic, excessive helium production rates, so care must be taken when interpreting their significance for fusion conditions.

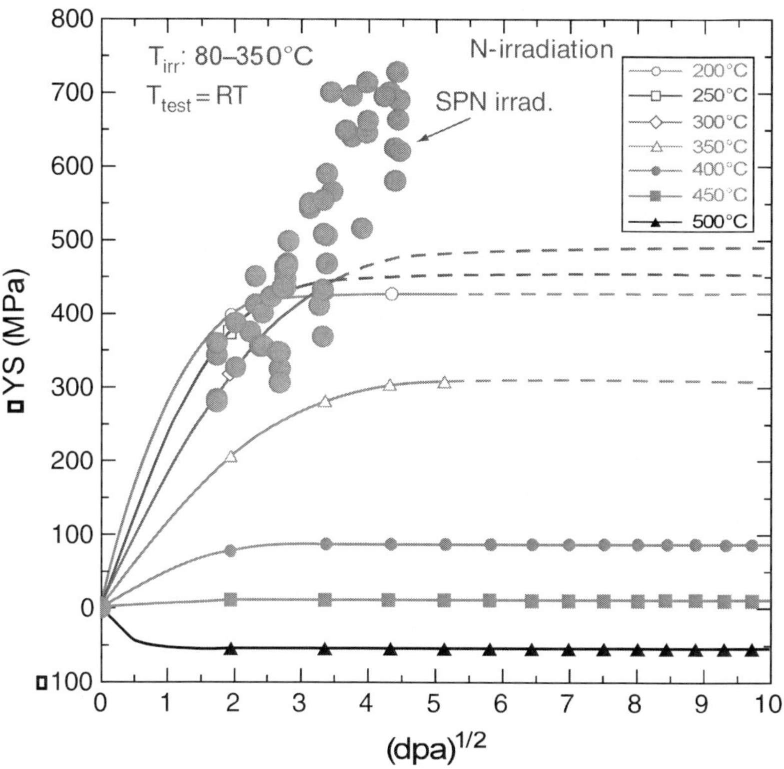

Figure 15. Model calculations (lines) of yield stress change versus neutron dose for ferritic/martensitic steels compared to hardening induced by irradiation in a mixed spectrum of high energy proton and spallation neutrons (SPN). The model calculations reflect the effects of displacement damage only while the SPN data include displacement damage and high levels of helium (~100 appm He/dpa) [68].

The upper operating temperature limit for RAF/M steels is about 550°C, which is controlled by the high-temperature stability of the tempered martensitic substructure, and the thermal creep strength [69, 70]. Thermal annealing of RAF/M steels show that they soften at temperatures ≥ 600°C [69] and this softening is accompanied by thermal embrittlement as indicated by an upward shift in the ductile-to-brittle transition temperature. Introduction of helium at high-temperatures could adversely impact the creep rupture strength of RAF/M steels if helium is able to collect at grain boundaries and enhance creep cavitation. There is little information on the potential severity of this effect due to the difficulty of conducting experiments on RAF/M steels at high-temperature, under stress and with fusion-relevant helium generation.

At intermediate irradiation temperatures, ~30 to 60% of the absolute melting temperature, dimensional instabilities associated with irradiation creep and swelling can significantly impact the operating envelope. In the absence of helium RAF/M steels exhibit excellent resistance to swelling up to doses around 200 dpa [66], but very little is known about the swelling or irradiation creep characteristics of RAF/M steels when large amounts of helium is present. Fusion-relevant levels of helium can increase the amount of swelling, but the maximum tolerable dose to avoid exceeding design margins is unknown [71].

Recent reviews of the state of RAF/M steel development have summarized what remains to be done to determine if these steels will be suitable for long-term fusion power plant applications [69, 71-74]. One conclusion from these reviews is that more work is needed to develop and qualify the manufacturing and joining technologies needed for future fusion devices. Most importantly for a next-step device and DEMO there is a strong need to fully quantify the effects of high levels of transmutation products such as helium and hydrogen on mechanical properties over the entire anticipated operating temperature window. Other critical issues include: further detailed assessment of irradiation-induced changes in mechanical properties; more work on fatigue and creep-fatigue mechanical property characterization and high-temperature design rule development; further exploration of RAF/M steel compatibility with potential coolants, purge gases and breeding materials such as flowing lead-lithium and solid ceramic breeding materials, also including the possible development of corrosion and/or permeation barrier coatings; definition of the maximum allowable helium concentration for repair welding; development of a new generation of RAF/M that does not require post weld heat treatment; and development of nondestructive examination techniques and procedures for flaw detection and sizing.

4.3.2. Nanostructured Ferritic Alloys

Nanostructured ferritic alloys (NFA) are a new class of material that offers the potential to operate at considerably higher temperatures than RAF/M steels and may be much more radiation tolerant. Nanostructured ferritic alloys are Fe-Cr alloys that derive their unique properties from the presence of an ultrahigh density of Y-Ti-O rich particles. These materials typically contain ≥ 12%Cr along with small additions of yttrium, titanium and oxygen [75]. They are different from oxide dispersion strengthened (ODS) steels, which are alloyed with C and often about 9%Cr. ODS steels are usually martensitic and generally contain coarser-scale oxides [75].

The high number density (~5 x 10^{23} m^3) of small (~2-5 nm diameter) Y-Ti-O particles efficiently impedes glide and climb of dislocations, which gives rise to exceptional high-temperature strength, as shown in Figure 16. The plot in Figure 16 shows a Larsen-Miller parameter plot of the creep-rupture strength of several NFAs compared to conventional RAF/M steels F82H and EUROFER. It is evident from these data that NFAs possess considerably better high-temperature strength than RAF/M steel. The upper operating temperature limit for NFAs is estimated to be in the range ~750 to 800°C, but a potential issue for long-term, high-dose applications might be embrittlement associated with radiation-induced precipitation of the undesirable Cr-rich alpha prime phase [71].

The nanometer-scale oxide particles also impart improved radiation resistance by providing vacancy-interstitial recombination sites that can mitigate displacement damage, and they act as preferred sinks to trap helium in small gas bubbles. Recent research shows that when particle sink strengths exceed ~$10^{16}/m^2$, radiation-induced hardening is decreased [66]. Distributing the helium amongst a high number density of small bubbles ameliorates several helium-related degradation phenomena. Significant void swelling is avoided by keeping helium bubbles below the critical size, which prevents conversion to rapidly growing voids. Trapping of nearly all the helium at the matrix Y-Ti-O particles shields the grain boundaries from acquiring large amounts of helium and causing premature failure by grain boundary creep cavitation.

Development of NFAs for fusion applications is in the very early stages [80]. While these materials possess many attractive features there are significant challenges that must be resolved before practical applications are feasible. Odette, Alinger and Wirth recently summarized some of the more important technical issues that must be addressed [67] such as: the high cost of mechanical alloying compared to more conventional melt-processing methods;

the difficulty of joining to obtain properties similar to the base material; fabrication of product forms with isotropic microstructures and properties; alloy homogeneity; relatively low fracture toughness for unirradiated material; lack of a robust database on many critical properties such as creep rates and rupture times, fatigue and creep-fatigue, fatigue crack growth, and corrosion; limited characterization of the effects of irradiation on microstructure and property evolution; and basic understanding of Y-Ti-O particle composition and structure, and how to control these variables to achieve optimal properties.

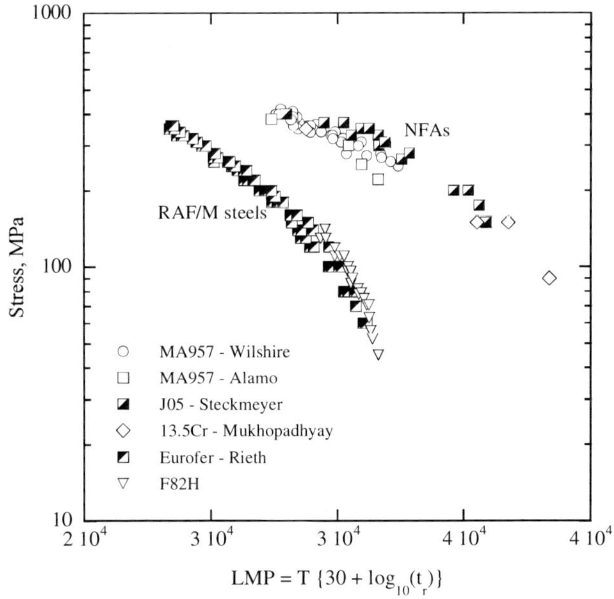

Figure 16. Larson-Miller parameter plot (C=30) comparing the creep-rupture strength of typical NFAs (MA957, J05, and 13.5Cr) with conventional RAF/M steels Eurofer and F82H [76-79].

4.3.3. Tungsten

Tungsten alloys have received considerable attention as promising materials for plasma facing component (PFC) applications for next-step fusion devices because they are the only realistic solid material for helium-cooled divertor applications due to their excellent thermo-physical properties. The main advantages of tungsten base alloys include; high melting point, good thermal conductivity, low sputtering and erosion yield, high strength, low thermal expansion, and high resistance to swelling.

Using tungsten for structural applications is very challenging because of its inherent brittleness. The fracture toughness is low and the ductile-to-brittle transition temperature of pure polycrystalline tungsten is relatively high. The fracture toughness data collected in Figure 17 illustrate that tungsten is very brittle below temperatures of about 600°C [81]. It should be noted that even at 600°C the median fracture toughness is only about 20 MPa-$m^{1/2}$, which does not provide substantial crack tolerance.

The toughness of tungsten can be improved by alloying with rhenium or by severe plastic deformation to very high strains to produce nanometer-size grains and high dislocation densities. Unfortunately neither of these approaches appears to be suitable for fabrication of tungsten components for fusion because rhenium is expensive and creates the potential for radiation-induced precipitation of embrittling phases, and fabrication of large components would be difficult to accomplish by the methods for producing severe deformation. Incorporation of particle dispersions such as La_2O_3, Y_2O_3 or TiC in tungsten may improve the mechanical properties by greatly refining the grain size and also increasing the recrystallization temperature, but precise control of the particle dispersions is required to effectively pin the grain boundaries and prevent grain growth during service. The challenges associated with alloying and preparation of nano-grain tungsten has motivated research on tungsten composites that utilize engineered reinforcement architectures for improved toughness. In principle this is achieved by incorporating ductile phases or strong reinforcing fibers that bridge a growing crack and lower the effective stress intensity at the crack tip.

Control of interstitial impurities is vital for obtaining acceptable mechanical properties of tungsten. Low solubility impurities segregate to grain boundaries reducing grain boundary cohesive energy, which can cause embrittlement. The concentration of grain boundary impurities can be kept low if the grain boundary area is large or, conversely, the grain size is small. The propensity for tungsten to recrystallize at high-temperature limits the upper use temperature. The effect of recrystallization is to increase the grain size and grain boundary interstitial impurity concentration, and lower the fracture toughness. The recrystallization temperature depends on composition and processing history but could be as low as ~1000°C [71].

While the neutron irradiation effects database on tungsten is limited, the available data shows that tungsten hardens and embrittles when irradiated at temperatures below about 35% of the melting point [82]. The peak neutron wall load and gas production for tungsten in a typical magnetic confinement fusion power system is considerably lower (by a factor of ~4) than for iron base alloys [63, 83], but the production of solid transmutation elements is greater in W [83, 84]. The principal solid transmutation products are rhenium, tantalum, and osmium. This transmutation process is quite sensitive to the location of W and its end-of-life fluence. Under ARIES power plant operating conditions, only 1% of the divertor W transmutes into Re, Ta, and Os at a typical 3-4 MWy/m^2 fluence. Such a low transmutation level may not affect the mechanical and physical properties of a tungsten divertor structure. However, tungsten armor attached to the first wall of a ceramic breeder blanket will experience twice as much transmutation for the same fluence because of the much softer neutron spectrum at the blanket surface, but the change in the armor properties may not be critical to the first wall [84]. Interestingly, the solid transmutation rates in fission reactors are much greater than for a fusion plant by about 5 to 50 times [83], which underscores the need for a fusion-relevant neutron source to properly assess irradiation effects in tungsten.

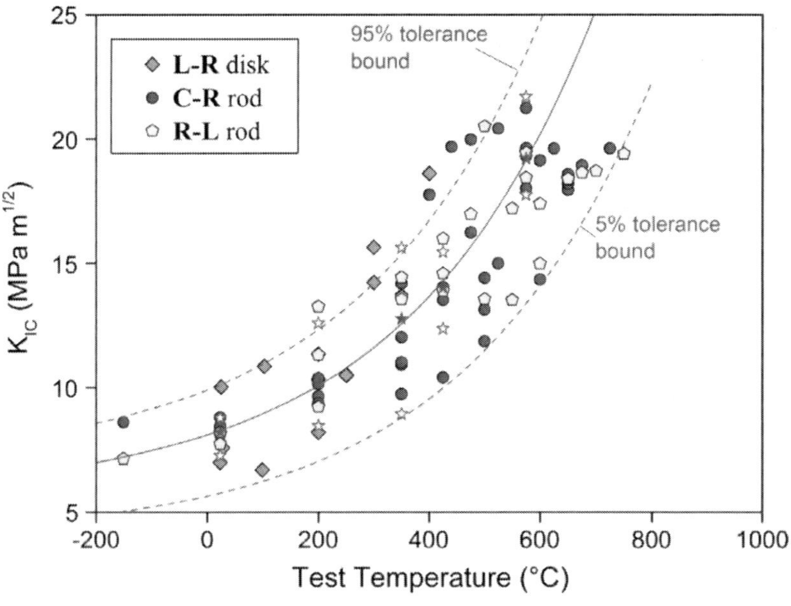

Figure 17. Temperature dependence of the fracture toughness of polycrystalline tungsten. The solid line is a fit to the median fracture toughness data using a Master Curve type function and the dashed lines are 95% and 5% tolerance bounds [81].

Three key factors must be considered when developing tungsten base divertor components for fusion including: 1) tungsten and its alloys possess fundamentally low cleavage toughness and impurity mediated low grain boundary cohesive energy, 2) the behavior of tungsten alloys in PFC structures will be determined not only by the material itself, but also the processing and fabrication procedures used to make components, and 3) tungsten and its alloys will be part of a complex system that is highly loaded, both thermally and mechanically, and that suffer a wide variety of degradation mechanisms. There are a number of fundamental critical issues that must be resolved to successfully develop and qualify tungsten for divertor applications including improving the creep strength, fracture toughness, microstructural stability, low and high cycle fatigue behavior, oxidation resistance, and the effects of neutron irradiation. Furthermore, there is a need to define the life-limiting mechanisms for tungsten structures and develop/approve design rules and codes for brittle materials.

Because of numerous performance concerns for tungsten, fusion researchers began to identify alternate high-Z materials for PFC (such as Zr, Nb, Mo, Ta, and Hf) and examine their activation characteristics, sputter erosion/redeposition properties, and plasma transient response. The initial results are encouraging, showing recycling capability and acceptable sputtering erosion/redeposition performances. More work is needed to advance the qualification of such materials for divertor applications [85].

4.4. Critical Research Needs to Develop Damage Tolerant Structural Materials

Overcoming the performance challenges of materials and structures in the fusion energy environment is a significant challenge, but yet it is critically important to realize the promise of fusion as a practical energy source. Development of materials, components and structures for any complex engineered system generally occurs in a series of steps, proceeding from relatively simple single-variable experiments to very complex, fully integrated, multiple-variable tests. The first step on the development path begins with screening experiments that are performed under carefully controlled, non-nuclear laboratory conditions. These experiments establish the basic mechanical and physical properties, chemical compatibility, and fabrication and joining technologies of candidate materials. If satisfactory results are obtained, more complex experiments are performed in order to identify materials that perform well in partially integrated or multiple-variable tests. The final stage of material development involves fully integrated experiments with full-scale components in prototypical test environments combining the appropriate nuclear, chemical, thermo-mechanical and magnetic fields necessary to establish lifetime and performance limits. The ultimate goal is to develop an extensive material database, constitutive equations, and physics-based models to describe all aspects of material behavior required for fusion power system design, construction, licensing and operation.

It has long been recognized that a fusion-relevant neutron source is one of the most important facilities needed for evaluating nuclear degradation of fusion structural materials [71]. Recent fusion materials research and development efforts have demonstrated that RAF/M steels and silicon carbide composites possess acceptable radiation resistance to neutron doses of order ~20 dpa, but the performance of these materials with accompanying high levels of gaseous and solid transmutants in a fusion neutron spectrum has not been established. Thus, a fusion relevant neutron source remains a critically important facility for determining the feasibility of generating fusion power and for developing a materials qualification database to satisfy regulatory authorities for licensing applications.

Zinkle and Möslang have carefully evaluated the facility options for acquiring the needed irradiation effects information [62]. In their study a broad range of irradiation facility options were considered accounting for the influence of key irradiation variables such as the primary knock-on atom spectrum, damage rate, pulsed irradiation effects and nuclear transmutations. A similar study was performed in the European Union [23]. These studies concluded that ion accelerators, fission reactors, and computational modeling would continue to play an indispensible role in the development of fusion materials. Spallation neutron sources were judged to be potentially useful for performing irradiation studies to relatively low damage levels (~10 dpa), but not to intermediate to high fluence levels. The major concern for spallation sources is the high-energy neutron tail that causes a considerable increase in gaseous and solid transmutation production that could significantly alter microstructural evolution and corresponding properties [62]. Plasma-based sources were also analyzed and found to possess attractive features such as near prototypic neutron spectra and the potential for large irradiation volume. However, such facilities, when compared to other options, were judged to be less attractive due to their projected high cost, long time for design and construction, and likely low availability [62]. The most attractive option for performing dedicated irradiations of fusion materials was determined to be an accelerator-based D–Li deuteron stripping source such as the International Fusion Materials Irradiation Facility [86], which was based on factors such as versatility and cost considerations.

It is clear that the capability to perform irradiation experiments under a variety of conditions (fission reactors, ion beams, spallation sources, etc.) is essential for finding the most promising materials and specimen geometries for further irradiation testing in the limited irradiation space that will be available in a fusion relevant neutron source and for developing robust models of materials behavior.

In the U.S. fusion program, the Fusion Nuclear Science Facility (FNSF) is currently under development. A material-testing module (MTM) will be included in FNSF to test large specimens of future generations of materials (RAF/M, W, blanket, divertor, magnets, etc.) and develop a comprehensive multi-materials database for a cumulative damage up to 125 dpa. The most important attributes for MTM would be a much larger high-flux irradiation volume (compared to the 10-500 ml range available in other irradiation sources), the opportunity to more fully understand materials behavior in a multi-effect environment (14 MeV neutrons, temperature gradient, pressure, stresses, magnetic field, etc.), and the evaluation and validation materials performance in a fusion-relevant environment [29].

Finally, while an intense fusion-relevant neutron source is clearly the overarching need for enabling progress towards a fusion DEMO, it is also apparent that critical materials degradation phenomena, including corrosion and high temperature deformation and fracture processes occur even in the absence of intense, high-energy neutron bombardment. Thus, a number of non-nuclear facilities will be needed to test and qualify materials, components and structures for the development of test blanket modules, next-step plasma devices and DEMO. The role of such facilities would be to investigate the potential for synergistic effects that are not revealed in simpler single-variable experiments or limited multiple-variable studies. These facilities would provide test beds for evaluation of operational procedures, transient event mitigation, nondestructive inspection techniques, and repair/maintenance procedures.

5. Need for Advanced Manufacturing Techniques

As ITER advances the knowledge and experience database for magnetic confinement of large fusion concepts, principally in the area of understanding and controlling very large plasmas, the world's fusion literate nations are now planning on how to migrate this knowledge into future large test facilities, demonstration plants and, ultimately, power plants. The next technological steps are to understand, develop and build the technologies and systems needed to achieve the required capabilities. This section addresses what new manufacturing technologies may enable these designs at an affordable cost.

Many of the other fusion-literate nations are proceeding on the inherent assumption of using currently available technologies to manufacture and fabricate their design concepts using the existing or near-term materials, processes and fabrication techniques with minimal emphasis on achieving a competitive energy source. Although the U.S. has not embraced or endorsed any official fusion energy plan, its past and current domestic emphasis is to adopt a more deliberate approach that emphasizes achieving the long term goal of higher efficiency plants that can competitively compete with existing and planned energy sources. This U.S. approach emphasizes a power core that operates in a higher temperature regime, is inherently safer, more environmentally attractive than other energy sources, and favorably competes economically with envisioned future energy sources.

This U.S. approach entails the use of new higher capability materials, relatively complex geometries and highly reliable subsystems and systems that can enable these ambitious goals. This chapter will explore how advanced manufacturing technologies may address these challenging needs.

5.1. Significantly Lower Cost Manufacturing

Affordability is an essential goal in the design and manufacture of any element of the facility to achieve the end goal of a reasonably priced project and, ultimately, the competitiveness of the facility. Thus, the fusion program must embrace cost-effective fabrication techniques that can build complex geometries with emerging material technologies to achieve a low-cost approach.

Basic fabrication materials have been used for some time and their unit costs are well known and are not likely to significantly change in the future. Currently, advanced nuclear materials with higher capabilities employ additional processing to achieve their augmented properties – thus, higher costs. The next material advancements will likely involve nano-manufactured materials. When this technology is perfected, this may enable tailored materials specifically designed for the component requirements (strength, ductility, heat transmission, radiation tolerance, low activation, etc.). The recent advances in 3-D printing of metals will likely enable a low cost capability to manufacture tailored materials and components for nuclear applications and achieve significant advances in properties, design-to-fabrication times, part and subsystems reliability (fewer parts) and lower capital and operational cost.

5.2. Additive Manufacturing Technologies

The prior subsection introduced the concept of 3-D printing as a method to fabricate complex geometries. This 3-D printing technology started around 1984 with the advent of stereo-lithography, a printing process that enables a tangible 3-D object to be created from digital data. Since then, the process has evolved using various forms of plastic and then metals to create objects directly from the digital data without any pre-forms, castings or prototypes. Now, many energy sources are used as the method to melt and place the material per the digital data.

Historically, parts are largely created starting from a large block of material and slowly the material is removed to reveal the finished part – akin to a sculptor starting with a marble block and finishing with a wonderful or artistic statue. This is generally referred to as “subtractive manufacturing”. On the other hand, this 3-D printing process is the exact opposite as it successively builds the geometry and is referred to as “additive manufacturing.”

As the 3-D printing capability evolves, it is expected that the product strength could increase, tolerances improve, surface finishes might require less or no processing, and the fabrication time and cost goes down. The biggest savings is the much lower material and labor costs to manufacture the part. This technology is commonly equated to the development of the printing press and the industrial revolution - it is that revolutionary!

Exploring the ability to generate complex parts, this process can produce very intricate parts, including interior passages, coolant channels and strengthening ribs (plus many other features, perhaps too small to see.) A decade ago, an additive manufactured wing spar was flown in a fighter aircraft, passing all flight-worthiness tests. In addition to being able to quickly reproduce complex parts, the process has the ability to seamlessly and gradually transition from one material to another. This would enable directly building entire subassemblies and assemblies without having to physically join them (no joint seams to leak, thus higher reliability). Also if the part needs increased strength or thermal/electrical conductivity in a certain orientation or element, this is easily accommodated.

If the material needs to be strengthened or enhanced by including special materials, e.g., thermal or electrical conductive materials, these additional materials could be precisely positioned in the material matrix. Sandwich materials could easily be made with internal cooling channels and manifolds all in one integral part. Even new classes of materials could be engineered and created.

The next step is nano-fabrication. Per the Nanobot.info page (http://www.nanobot.info/), *"Basic nanomachines are already in use. Nanobots will be the next generation of nanomachines. Advanced nanobots will be able to sense and adapt to environmental stimuli such as heat, light, sounds, surface textures, and chemicals; perform complex calculations; move, communicate, and work together; conduct molecular assembly; and, to some extent, repair or even replicate themselves. The extreme concept of nanotechnology is the "bottoms up" creation of virtually any material or object by assembling one atom at a time."* This technology might be extremely advantageous to enable in-situ repair and replacement of the damaged or life-limited components inside the fusion power core. It is likely this may not be ready in the time frame for the next big technology test facility, but if this capability proves to be useful and advances sufficiently in time to be implemented, this indeed could be a "game changer".

In general, it is not conceived that advancing the additive manufacturing and nano-fabrication would be funded and developed within any country's fusion budget. Rather, we must keep apprised of these developments and perhaps influence the end applications. The exception might be the fabrication of fusion-unique materials and parts for nuclear systems. We need to stay abreast of this exciting and rapidly evolving manufacturing revolution.

6. Tailoring of Maintenance Scheme to Plant Needs

Within the fusion power core, the intense neutrons released in the D-T fusion process causes the in-vessel components to transmute into "mildly" radioactive isotopes. Although these radiation levels are short lived compared to those created during the fission process, the fusion irradiated materials are sufficiently activated to not allow any human exposure inside the bioshield barrier. Also, all material properties will be degraded at their end of life to the point where most internal components must be replaced. With these levels of radiation, all operations inside the bioshield must be accomplished completely by remote means.

6.1. Confirming the Facility Maintenance Approach

To assure that the remote maintenance system can function as required, ITER [48, 87, 88] and all subsequent large fusion facilities are planning on remotely assembling their power cores as a final test of the maintenance approach. Additionally, it is recommended that prior to committing to a particular approach and equipment, the complete power core and maintenance equipment be validated in a test facility with fusion relevant environment.

6.2. Need for High Plant Availability

One of the most important parameters for a facility is known as plant availability, defined as the time the facility can operate divided by the plant total time (up time / (up time + down time). ITER and other experimental fusion facilities are not concerned about the overall plant availability. Rather, they are most interested in the experimental availability or having the facility operational during test periods. Downtimes can be rather lengthy so long as the test periods are not disturbed by outages.

On the other hand, future fusion technology-related test facilities [20] will be required to run continuously for longer periods of time to accumulate valuable and necessary test data. Projecting this to the DEMO and the first power plant significantly raises the availability requirement (85-95%). Setting the availability goal this high requires extremely reliable systems, long-lived components and a highly efficient and expedient maintenance system that is completely integrated with the power core and facility design and arrangement.

6.3. The Need for a Highly Integrated Power Core, Facility and Maintenance Approach

Developing a maintainable power core and efficient maintenance system has been a cornerstone of all conceptual fusion demonstration and power plant design studies since the early 1970s with the advent of UWMAK-1 [89]. All the following fusion plant studies recognized the importance of maintaining the power core in a timely fashion, ultimately to achieve high plant availability to produce competitive electrical power. The plant architects and designers devised plans to quickly and efficiently remove and replace the power core elements. The maintenance approach, remote handling equipment and building layout must be completely compatible with the power core design to achieve the facility goals.

6.4. Bigger Modules Are Better

One of the common trends recognized by many of the engineers involved in conceptual designs for fusion power plants is that it is more beneficial for plant availability and reliability of the renovated power core to replace very large modules - the bigger the better. The largest tokamak power core slice that could be removed and replaced is a complete sector between each individual TF coils, typically 12 or 16 coils. This trend also holds for stellarators, but the stellarator TF coil geometry is so convoluted, that whole sectors could not be removed and smaller modules were chosen [90]. For a spherical tokamak, this concept was carried to the extreme as it was proposed to vertically replace the entire power core out the bottom of the plant [91]. The concept of large modular replacement to achieve quicker maintenance and improved reliability is a common trend in most high-technology products. It is faster and more reliable to remove and replace a complete article than it is to diagnose and repair smaller discrete elements in situ.

6.5. Which is Best, Horizontal or Vertical Access?

The early and current U.S. fusion plants embraced the approach of removing large sectors horizontally between each of the TF coil legs via large ports attached to the vacuum vessel. To create an adequately wide opening for removal, the horizontal approach requires the outer TF coil legs to be slightly increased in the radial direction. The advantage of this approach is that the sectors can be moved with a single radial motion into a large shielded cask for transport to the hot cell. The ability to remove any sector without disturbing its neighbors is quite an advantage for any unscheduled maintenance action. The detriment for this approach is that the sector is quite massive, even with the coolants removed.

Alternatively, Europe and some other countries have adopted a vertical maintenance approach to remove smaller segments (a subdivision of the sector) through openings between all or a few of the TF coil upper legs. Since the access between the upper TF coils is quite restricted as compared to the outboard midplane, the removed elements must be substantially narrower than the sector width. Another restriction is that the upper port location (size) usually requires either a very narrow segment or the segments are subdivided into inboard and outboard segments. With the single null divertor approach, the bottom divertor cassettes are usually removed through a lower port. Typically, one segment will be directly below the vertical port and can be lifted by an overhead crane out of the power core into a shielded cask for transport to the hot cell. The remaining segments must be move radially, toroidally and then vertically. These internal power core segment movements must be accomplished with additional in-vessel manipulators that use the lower divertor cassette rails. This maintenance procedure entails a lot of sequential moves with additional in-vessel equipment and disconnections and reconnections. When unscheduled failures occur or premature replacements are required, this approach may have to remove a substantial portion of the power core to reach the affected component, depending on the particular power core design.

The choice of horizontal or vertical maintenance approach depends on determining which technique offers the easiest, most expedient, most reliable and most flexible means to achieve the required plant availability for the facility as well as future power plants.

6.6. Precision Placement of Replaced Articles

With any fusion power core maintenance plan, a universal requirement is for all the plasma facing surfaces to be aligned with close tolerances to ensure the expansion gaps between adjacent sectors or segments are within the correct dimensions. The caveat to that prior statement is that the installation procedures must account for local erosion and swelling/distortion of the mating elements, perhaps enabled by special localized diagnostics. These requirements would suggest the use of automated remote handling equipment to accurately repeat article placement. It is also likely that an in-vessel inspection capability will be necessary to determine the correct placement.

6.7. Maintenance R&D Needs

The design and development of the maintenance plan and associated equipment cannot be accomplished independently of the fusion power core design as the maintenance approach and the power core are intimately interconnected. Prior to committing to a final design, it would be advisable to construct and operate a non-nuclear power core and maintenance validation facility at full scale and fidelity to assure all operations will work as intended. This facility could also evolve the maintenance procedures and improve power core elements and interfaces. Another R&D effort would involve developing radiation resistant (tolerant) maintenance and diagnostic equipment for the environment inside the power core and inside the hot cell and waste handling buildings.

7. Handling and Processing of Fusion Radioactive Materials

Fusion has long been envisioned as possessing an inherent advantage for having a relatively benign environmental impact, mainly due to the absence of high level waste (HLW) generation. However, fusion will likely generate a sizable amount of mildly radioactive materials. Such a potential problem has been overlooked in early fusion studies and/or relegated to the back-end as only a disposal issue in near-surface, low-level waste (LLW) repositories, adopting the preferred radioactive waste (radwaste) management approach of the 1960s. In fact, existing disposal sites are not sized to accommodate the large volume of fusion LLW generated during operation and after decommissioning. To put matters into perspective, Figure 18 compares the power core volumes of ITER [48], the advanced ARIES power plants (ARIES-ACT-1&2) [6] and the European Power Plant Conceptual Study (PPCS) [5] to ESBWR (Economic Simplified Boiling Water Reactor) – a Gen-III$^+$ advanced fission reactor [92]. Clearly, fusion generates much larger quantities of radioactive materials compared to fission.

The shallow land burial is not a viable option for fusion because of the large radwaste. Thus, it is essential to consider a more environmentally attractive framework to keep the volume of fusion LLW to a minimum. This requirement is important to the future of fusion energy. Concerns about the environment, radwaste burden for future generations, lack of geological repositories, and high disposal cost directed our attention to recycling of the radioactive materials (for reuse within the nuclear industry) and clearance (the unconditional release to the commercial market if materials contain traces of radioactivity).

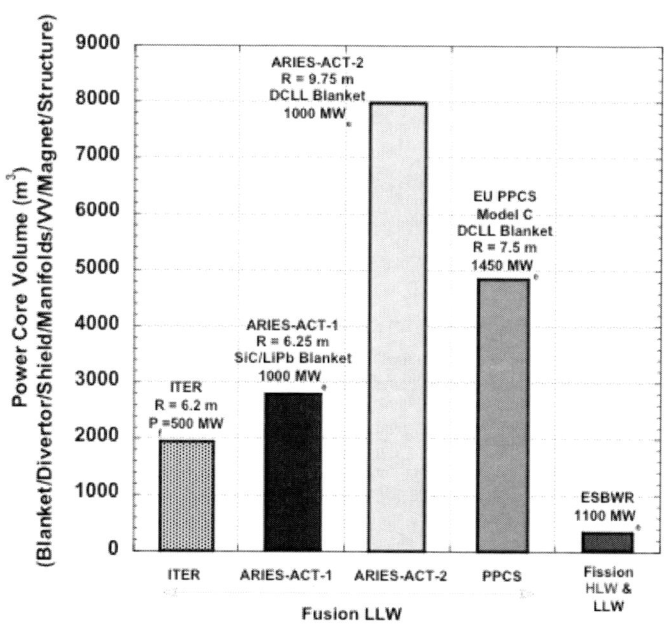

Figure 18. Comparison of radioactive waste from power core of fusion and fission designs (actual volumes of components; not compacted; no replacement).

In the 1980s and 1990s, fusion researchers have investigated the recycling and clearance options. More recently, these approaches became more technically feasible with the development of advanced radiation-hardened remote handling (RH) tools that can process the highly radioactive materials with up to 10,000 Sv/h doses [93, 94], along with the introduction of the clearance category for slightly radioactive materials by national and international nuclear agencies [95, 96]. Such recent advances encouraged many fusion designers in the U.S. and throughout the world [97-103] to apply the recycling and clearance approaches to all fusion components that are subject to extreme radiation levels: very high level near the plasma and very low level at the biological shield (bioshield). If the design incorporates the recycling and clearance requirements from the start, fusion will eventually reach the ultimate goal of radwaste-free energy source.

Numerous fusion studies indicated the recycling/clearance approach is relatively easy to envision and apply from a science perspective. To assess this argument, all three scenarios (disposal, recycling, and clearance) were considered for the most recent ARIES-ACT-2 design [28] in the ARIES power plant series.

7.1. Assessing Disposal Issues and Needs

The ARIES-ACT-2 waste disposal rating (WDR) is evaluated for fully compacted components using the waste disposal limits developed by Nuclear Regulatory Commission (NRC) 10CFR61 [104] and Fetter [105]. The NRC waste classification is based largely on radionuclides that are produced in fission reactors, hospitals, research laboratories, and food irradiation facilities. In the early 1990s, Fetter and others [105] performed analyses to determine the Class C specific activity limits for all long-lived radionuclides of interest to fusion using a methodology similar to that of 10CFR61 [104]. Although Fetter's calculations carry no regulatory endorsement, they are useful because they include fusion-specific isotopes. All ARIES components should meet both NRC and Fetter's limits until the NRC develops official guidelines for fusion waste. Also, the WDR is evaluated at 100 y after shutdown, allowing the short-lived radionuclides to decay. A WDR < 1 means LLW and WDR > 1 means HLW.

There is a need to strictly limit the undesirable Nb and Mo impurities in fusion materials to below 0.5 and 5 wppm, respectively, to prevent the generation of HLW. Such an impurity control mandate would allow all in-vessel components to achieve the Class C LLW classification. The ex-vessel components (shield and magnets) will be less radioactive and would qualify as Class A LLW – the least hazardous type of waste. Excluding the clearable components (cryostat and bioshield), ~40% of the ARIES-ACT-2 waste (FW, blanket, structural ring, vacuum vessel (VV), and divertor) would qualify as Class C LLW. The dominant radioisotope remaining at 100 y after shutdown is ^{94}Nb (from the 0.5 wppm Nb impurity in the RAF/M F82H steel structure). The remaining ~60% of the radioactive waste (LT shield and magnet) would be classified as Class A LLW.

Several critical issues for the disposal option are identified based on the assessment of disposal situation in the U.S. These are listed below along with specific needs for the fusion community:

Disposal issues:

- High disposal cost that continues to increase (for preparation, characterization, packaging, interim storage, transportation, licensing, and disposal)
- No commercial HLW repositories exist in the U.S.
- Limited capacity of existing LLW repositories
- Political difficulty of siting new repositories limits their capacity
- Prediction of repositories' conditions for long time into future
- Radwaste burden for future generations.

Disposal needs:

- Revised LLW activity limits for fusion radioisotopes issued by NRC
- Official activity limits for Greater Than Class C fusion radwaste
- Repositories designed for T-containing materials
- Reversible disposal process and retrievable waste (to gain public acceptance and ease licensing)
- Large capacity and low-cost interim storage with decay heat removal capability.

7.2. Assessing Recycling Issues and Needs

The technical feasibility of recycling is dependent on the dose rate to the RH and processing equipment. Essentially, the dose determines the RH needs (hands-on, conventional, or advanced tools to handle the radioactive components) and the interim storage period necessary to meet the dose limit. Advanced RH equipment has been used in the nuclear industry, in hot cells and reprocessing plants, and in spent fuel facilities. While the fission processes may have no direct relevance to fusion, their success gives confidence that advanced RH and processing techniques could be developed to handle high doses (> 10,000 Sv/h) for the recycling of fusion materials. Beside the recycling dose, other important criteria include the decay heat level during reprocessing, recycling of tritium-containing materials, physical properties of recycled products, and economics of fabricating complex shapes remotely.

The recycling dose for ARIES-ACT-2 components is shown in Figure 19. The variation of dose with time shows a strong dependence on location within the power core. The highly irradiated components can potentially be recycled in less than a year after shutdown with advanced RH equipment that can handle tolerate 10,000 Sv/h or more. The FW surrounding the plasma is an integral part of the blanket. It is shown in Figure 19 as a separate component to illustrate it is the major contributor to RH equipment. However, the dose when averaged over the complete FW/blanket subsytem would be an order of magnitude lower. Mn-54 (from Fe) is the main contributor to the dose of FS-based components (FW, blanket, structural ring, and VV) at early cooling periods (<10 y), while impurities have no significant contribution to the recycling dose during such a period. Storing the FW/blanket temporarily for several years would help drop the dose by a few orders of magnitude before recycling. The outermost power core components (cryostat and bioshield) would have very low radioactivity levels and could be recycled with hands-on following a specific cooling period. No significant dose build-up is expected due to the reuse of materials after numerous life cycles based on previous multiuse analysis performed for the ARIES divertor [102].

As the recycling approach plays an essential role in minimizing the volume of radwaste assigned for geologic disposal, it should be pursued despite the lack of detail on how to implement it at the present time. Based on known technologies involved in the recycling process, several key issues are identified for the fusion community to examine with a dedicated R&D program:

Recycling issues:

- Separation of various activated solid materials from complex components
- Radiochemical or isotopic separation processes for some materials
- Properties of recycled materials? Any structural role? Reuse as filler?

- Treatment and remote re-fabrication of recycled materials
- Radiotoxicity and radioisotope buildup and release by subsequent reuse
- Inspection of parts made of recycled materials
- Handling of T-containing materials during recycling
- Management of secondary waste. Any materials for disposal? Volume? Radwaste level?
- Energy demand for recycling process
- Cost of recycled materials
- Recycling plant capacity and support ratio.

Recycling needs:

- Radiation-resistant remote handling and processing equipment
- Reversible assembling process of components and constituents (to ease separation of materials after use)
- Efficient detritiation system to remove tritium before recycling
- Large and low-cost interim storage with decay heat removal capacity
- Nuclear industry should accept recycled materials
- Recycling infrastructure.

7.3. Assessing Clearance Issues and Needs

The majority of nuclear waste (essentially the bioshield) contains traces of radioactive nuclides that represent no risk to the public safety and health. Over the past several decades, researchers in the nuclear field have attempted to issue policy statements to deregulate such materials with low concentrations of radioactive contamination. If this effort succeeds, the clearable materials will not be subject to regulatory control, can be handled as if they are no longer radioactive, and be unrestrictedly recycled into consumer products (tables, chairs, tools, building and road materials, etc.). In 2003, the NRC declared materials with low concentrations of radioactivity can be deregulated and issued the NUREG-1640 documents [95] to obtain the clearance indices (CI) for the 115 radioisotopes that can be found in four types of clearable materials: steel, copper, and aluminum scrap, and concrete rubble. In 2004, the International Atomic Energy Agency (IAEA) published clearance standards [96] for 257 radionuclides, taking into account the US NUREG-1640 evaluation [95].

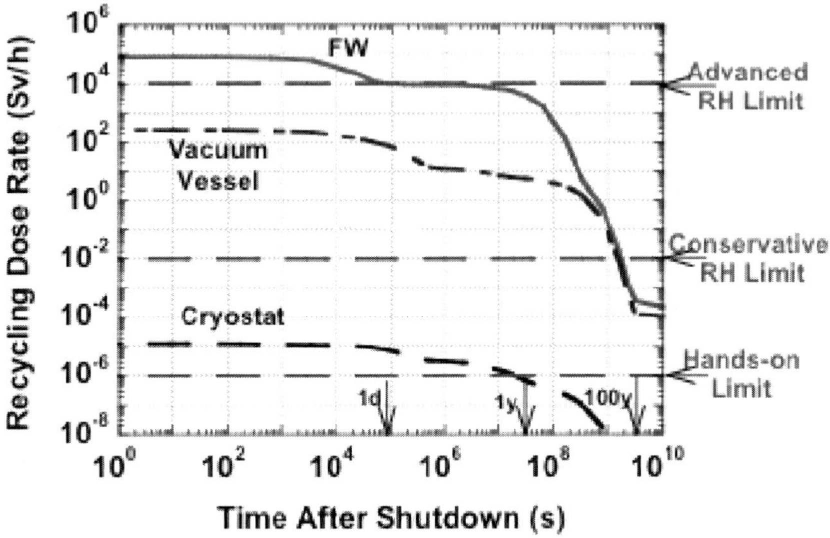

Figure 19. Recycling dose rate to RH equipment for selected ARIES-ACT-2 components [28].

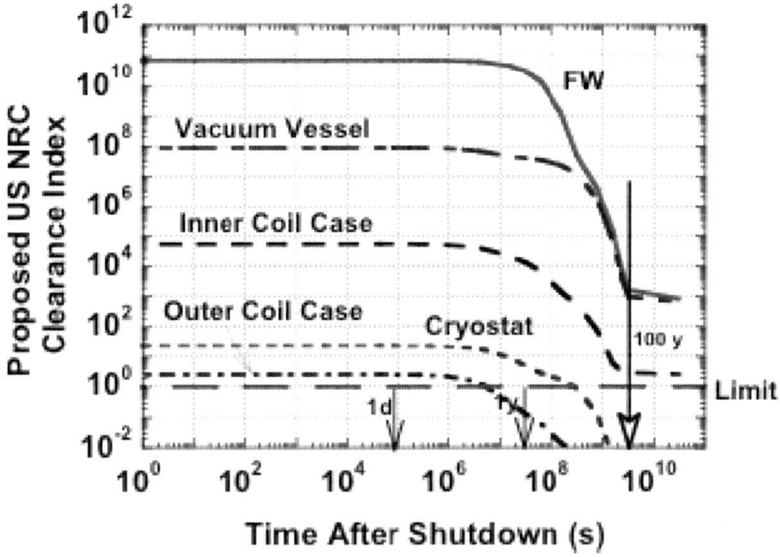

Figure 20. Variation of CI with time after shutdown for selected ARIES-ACT-2 OB components [28].

The NRC guidelines are referred to as the proposed U.S. CI limits since the NRC has not yet issued an official policy on the unconditional release of materials. The CI for a clearable material is calculated as the ratio of the activity (in Bq/g) of the individual radioisotope to the allowable clearance limit summed over all radioisotopes. A material qualifies for clearance if the CI drops below one at any time during a defined storage period following replacement or decommissioning. For components with multiple materials, it is essential to segregate and re-evaluate the CIs for constituents. Sizable components, such as the 2-m thick bioshield, should be segmented into small layers with the CI for each layer reexamined.

For ARIES-ACT-2, the CIs for all in-vessel components exceed unity by a wide margin even after an extended period of 100 y (refer to Figure 20). Nb-94 (from Nb impurity) is the main contributor to the CI of FS-based materials at 100 y. An OB VV made of pure iron would have a CI of ~7 at 100 y (from Co-60 generated by Fe). This means no matter how serious an attempt is to strictly control all impurities or alter the alloying elements of the ferritic steel, the VV (and LT shield) will never qualify for clearance, however it remains recyclable. Even though the magnet is well protected by the blanket, VV and shield, only its Cu stabilizer can be cleared in less than 100 y. The 2 m thick external concrete building (bioshield) that surrounds the tokamak represents the largest single component of the decommissioned materials. If adequately protected against streaming neutrons, the bioshield along with the cryostat qualifies for clearance, representing ~65% of the total volume of ARIES-ACT-2 radioactive materials.

The results indicate the ARIES-ACT-2 cryostat, bioshield, and some magnet constituents are clearable. It is unlikely that the missing fusion-related radioisotopes [106] from both U.S. NRC and IAEA CI evaluations could alter the conclusions. Nevertheless, efforts by the U.S. NRC, IAEA, and other organizations should continue to develop clearance standards for all radioisotopes of interest to fusion applications. The national and international organizations' efforts should be supported to convince industrial as well as environmental groups that clearance of slightly radioactive solids (that contain < 1% of background radiation) can be conducted safely with no risk to the public health. Other clearance issues and needs that require further assessment include:

Clearance issues:

- Discrepancies between proposed US-NRC and IAEA clearance standards

- Impact on clearance index prediction of missing radioisotopes (such as ^{10}Be, ^{26}Al, ^{32}Si, $^{91,92}Nb$, ^{98}Tc, ^{113m}Cd, ^{121m}Sn, ^{150}Eu, $^{157,158}Tb$, $^{163,166m}Ho$, ^{178n}Hf, $^{186m,187}Re$, ^{193}Pt, $^{208,210m,212}Bi$, and ^{209}Po)
- Radioisotope buildup and release by subsequent reuse.

Clearance needs:

- Official clearance limits issued by NRC
- Accurate measurements and reduction of impurities that deter clearance of in-vessel components
- Reversible assembly process of components and constituents
- Large capacity and low-cost interim storage facility
- Clearance infrastructure
- Clearance market.

7.4. Summary

The amount of fusion low-level waste is anticipated to be large, thus efforts to avoid the use of geological disposal is encouraged by the promotion of recycling and clearance of all components. The desire to reclaim valuable resources in the form of metal alloys and concrete rubble persists in all industries, not only nuclear. To make the recycling and clearance approaches a reality, major rethinking, education, and research should be developed and pursued in the nuclear field. In the meantime, a fusion program should be set up to accommodate the recycling/clearance strategy to embrace the promise of fusion energy production with low environmental impact. At present, the experience with recycling/clearance is limited, but will be augmented significantly by advances in fission reactor dismantling, spent fuel reprocessing, and bioshield clearance before fusion is committed to commercialization in the second half of the 21st century.

Conclusion

Significant technology challenges remain to be solved that will require substantial effort and investment to advance fusion from the present experimental phase to commercially viable power plant. ITER, the largest D-T fuelled tokamak experiment, will demonstrate in the 2020s the necessary technologies to achieve fusion ignition for very short operating times. Specifically, these technologies include low-temperature water-cooled shield (in place of a breeding blanket), vacuum vessel, and divertor; plasma fuelling, heating, and current drive systems; low-temperature cryogenic superconducting magnets; diagnostics; remote maintenance of small modules; water detritiation system; and geologic disposal of radioactive waste. These demonstrations will be very significant accomplishments; however these technologies must be further advanced and matured for DEMO and future power plants. In addition, other technology developments will be necessary for high-temperature FW and tritium breeding blankets, high heat flux divertors, advanced radiation-resistant structural materials, efficient tritium control and extraction systems, advanced and low-cost manufacturing processes, effective maintenance schemes, and recycling and clearance of radioactive materials. Achieving the necessary levels of reliability and maintainability for fusion components and subsystems are equally challenging tasks, requiring substantial testing, qualification and validation, far beyond what has been achieved so far. Testing and long-term operations in a relevant fusion environment with 14 MeV neutron source will be needed for licensing of DEMO and subsequent power plants. Given the large number of fusion components, subsystems and systems, this endeavor will require long time and substantial investment that tend to have a profound impact on the realization of fusion energy. Worldwide, the fusion roadmaps may take different approaches, depending on the degree of physics extrapolation beyond ITER, technology readiness, and economic competitiveness. Nevertheless, all pro-fusion countries agree on a common goal: fusion must be an economically competitive and environmentally attractive source of energy.

ACKNOWLEDGMENTS

The authors would like to acknowledge the support of this work by their institutions: the Fusion Technology Institute at the University of Wisconsin-Madison and the Pacific Northwest National Laboratory in the U.S., and the Karlsruhe Institute of Technology in Germany. The authors are very appreciative of the contributions and comments received from Dr. S. Malang (Germany) for Section 2.

REFERENCES

[1] Laila, A. El-Guebaly. (2011). "History and Evolution of Fusion Power Plant Studies: Past, Present, and Future Prospects." Chapter 6 in book: Nuclear Reactors, Nuclear Fusion and Fusion Engineering. A. Aasen and P. Olsson Editors. NOVA Science Publishers, Inc.: Hauppauge, New York, USA. ISBN: 978-1-60692-508-9 (2009) 217-271. Also published in E-Journal: *International Journal of Energy, Environment, and Economics*, Volume *18*, Issue 1-2. ISSN 1054-853X 115-167. https://www.novapublishers.com/catalog/product_info.php?products_id =8852.

[2] Najmabadi, F., et al., (1997). "Overview of ARIES-RS reversed shear power plant study," *Fusion Engineering and Design*, *38,* 3-25.

[3] Kolbasov, B. N. et al., (2008). "Russian concept for a DEMO-S demonstration fusion power reactor," *Fusion Engineering and Design*, *83*, 870-876.

[4] Raffray, A. R., et al., (2007). ''Advanced power core system for the ARIES-AT power plant,'' *Fusion Engineering and Design*, *82*, 217.

[5] Maisonnier, D. et al., (2005). "A conceptual study of commercial fusion power plants," Final Report of the European Fusion Power Plant Conceptual Study (PPCS), EFDA-RP-RE-5.0. Available at:http://www. efda.org/eu_fusion_programme/scientific_and_technical_publications. htm.

[6] Kessel, C. E., et al., (2015). "The ARIES Advanced and Conservative Tokamak (ACT) Power Plant Study," *Fusion Science and Technology*, *67*, No. 1, 1-21.

[7] Najmabadi, F. & The ARIES Team. (2003). "Spherical torus concept as power plants–the ARIES-ST study," *Fusion Engineering and Design*, *65*(2) 143-164.

[8] Wong, C. P. C., et al., (2004). "APEX advanced ferritic steel, Flibe self-cooled first wall and blanket design," *Journal of Nuclear Materials*, *329-333*, 1599-1604.

[9] Sagara, A., et al., (2014). "Helical reactor design FFHR-d1 and c1 for steady-state DEMO," *Fusion Engineering and Design*, *89*, 2114-2120.

[10] Giancarli, L., et al., (1998). "Development of the EU water-cooled Pb-17Li blanket," *Fusion Engineering and Design*, *39-40*, 639-644.

[11] Li Puma, A. et al., (2012). "Design and development of DEMO blanket concepts in Europe," First IAEA DEMO Program Workshop, UCLA, USA, October 15-18. http://advprojects.pppl.gov/Roadmapping/IAEADEMO/Presentations/Topic1_Pres/Li_Puma_IAEA_DEMO.pdf.
[12] Boccaccini, L.V. (2013). "Progress in EU blanket technology," *Fusion Science and Technology*, *64*, No. 3, 615-622.
[13] Aubert, J., et al., (2014). "Development of water cooled lithium lead blanket for DEMO," *Fusion Engineering and Design*, *89*, 1386-1391.
[14] Sardain, P., et al., (2006). "The European power plant conceptual study: helium-cooled lithium-lead reactor concept," *Fusion Engineering and Design*, *81*, 2673-2678.
[15] Aiello, G., et al., (2011). "HCLL TBM design status and development," *Fusion Engineering and Design*, *86*, 1444-1450.
[16] Aiello, G., et al., (2014). "Development of helium cooled lithium lead blanket for DEMO," *Fusion Engineering and Design*, *89*, 2129–2134.
[17] Wu, Y. (2011). "Overview of liquid lithium lead breeder blanket program in China," *Fusion Engineering and Design*, *86*, issues 9-11, 2343-2346.
[18] Norajitra, P., et al., (2001). "The EU advanced lead lithium blanket concept using SiC_f/SiC flow channel inserts as electrical and thermal insulators," *Fusion Engineering and Design*, *58-59*, 629-634.
[19] Raffray, A. R. et al., (2008). ''Engineering design and analysis of the ARIES-CS power plant," *Fusion Science and Technology*, *54*, 725.
[20] Kessel, C., et al., "The fusion nuclear science facility (FNSF), the critical step in the pathway to fusion energy," to be published in Fusion Science and Technology.
[21] Menard, J., et al., (2014). "Configuration ftudies for an ST-based Fusion nuclear science facility," Proceedings of 25th IAEA Fusion Energy Conference, Oct. 13-18, St. Petersburg, Russia.
[22] Sawan, M., et al., (2005) "Neutronics Assessment of Molten Salt Breeding Blanket Design Options," *Fusion Science & Technology*, *47*, No. 3, 510-517.
[23] Stork, D., et al., (2014). "Materials R&D for a timely DEMO: Key findings and recommendations of the EU Roadmap Materials Assessment Group," *Fusion Engineering and Design*, *89*, 1586-1594.

[24] Malang, S., et al., (2011). "Development of the lead lithium (DCLL) blanket concept,'' *Fusion Science and Technology*, *60*, 249.

[25] Sharafat, S., et al., (2009). ''Development status of a SiC-foam based flow channel insert for a U.S.-ITER DCLL TBM,'' *Fusion Science and Technology*, *56*, 883.

[26] Norajitra, P., et al., "Development of sandwich flow channel inserts for an EU DEMO dual coolant blanket concept," to be published in Fusion Science and Technology.

[27] Wang, X. R., et al., (2015). "ARIES-ACT2 DCLL power core design and engineering," *Fusion Science and Technology*, *67*, 193-219.

[28] El-Guebaly, L. & Mynsberge, L. (2014). "Neutronics characteristics, shielding system, activation and environmental aspects of ARIES-ACT-2 power plant," University of Wisconsin Fusion Technology Institute Report, UWFDM-1418. Available at: http://fti.neep.wisc.edu/pdf/fdm1418.pdf.

[29] El-Guebaly, L., Malang, S., Rowcliffe, A. & Waganer, L. "Blanket/materials testing strategy for FNSF and its breeding potential," to be published in Fusion Science and Technology.

[30] Malang, S., et al., (2009). "An example pathway to a fusion power plant system based on lead–lithium breeder: Comparison of the dual-coolant lead–lithium (DCLL) blanket with the helium-cooled lead–lithium (HCLL) concept as initial step," *Fusion Engineering and Design*, *84* 2145-2157.

[31] Ying, A. et al., (2012). "Status of ceramic breeder pebble bed thermo-mechanics R&D and impact on breeder material mechanical strength," Fusion Engineering and Design, *87*, 1130-1137.

[32] Knitter, R., Chaudhuri, P., Feng, Y. J. & Hoshino, Yu, I-K. (2013). "Recent developments of solid breeder fabrication," *Journal of Nuclear Materials*, *442*, 420–424.

[33] Chakin, V., Rolli, R., Moeslang, A., Klimenkov, M., Kolb, M., Vladimirov, P., Kurinskiy, P., Schneider, H-C. van Til, S., Magielsen, A. J. & Zmitko, M. (2013). "Tritium release and retention properties of highly neutron-irradiated beryllium pebbles from HIDOBE-01 experiment," *Journal of Nuclear Materials*, *442*, 483–489.

[34] Uchida, M., Kawamura, H., Uda, M. & Ito, Y. (2003). "Elementary development of beryllide pebble fabrication by rotating electrode method," *Fusion Engineering and Design*, *69*, 491-498.

[35] Dalle Donne, M., et al., (1991). "Demo-relevant test blanket for NET/ ITER. BOT helium cooled solid breeder blanket," Kernforschungszentrum Karlsruhe Report, KfK 4928 and 4929.

[36] Dalle Donne, M., Albrecht, H., Boccaccini, L. et al., (1998). "European helium cooled pebble bed blanket design of blanket test module to be tested in ITER," *Fusion Engineering and Design*, *39-40*, 825-833.

[37] Enoeda, M., Tanigawa, H., Hirose, T., Nakajima, M., Sato, S., Ochiai, C. Konno, K., Kawamura, Y., Hayashi, T. Yamanishi, T., Hoshino, T., Nakamichi, M., Tanigawa, H., Nishi, H., Suzuki, S., Ezato, K., Seki, Y. & Yokoyama, K. (2014). "R&D status on water cooled ceramic breeder blanket technology," *Fusion Engineering and Design*, *89*, 1131–1136.

[38] Feng, K. M., Zhang, G. S., Hu, G., Chen, Y. J., Feng, Y. J., Li, Z. X., Wang, P. H., Zhao, Z., Ye, X. F., Xiang, B., Zhang, L., Wang, Q. J. Cao, Q. X., Zhao, F. C., Wang, F., Liu, Y., Zhang, M. C. & Chinese HCCB TBM team. (2014). "New progress on design and R&D for solid breeder test blanket module in China," *Fusion Engineering and Design*, *89*, 1119-1125.

[39] Cho, S. Y., et al., (2014). "Design and R&D progress of Korean HCCR TBM," *Fusion Engineering and Design*, *89*, 1137-1143.

[40] Tavassoli, A-A. F., Diegele, E., Lindau, R., Luzginova, N. & Tanigawa, H. (2014). "Current status and recent research achievements in ferritic/ martensitic steels," *Journal of Nuclear Materials*, *455*, 269–276.

[41] Franza, F. (2013). "Tritium transport analysis in HCPB blanket with the FUS-TPC Code," KIT Scientific Report, 7642.

[42] Hermsmeyer, S., Malang, S., Fischer, U. & Gordeev, S. (2003). "Lay-out of the He-cooled solid breeder model B in the European power plant conceptual study," *Fusion Engineering and Design*, *69*, 281-287.

[43] Federici, G., et al., (2014). "Overview of EU DEMO design and R&D activities," *Fusion Engineering and Design*, *89*, 882-889.

[44] Liu, C., et al., (2011). "Nuclear analysis of DEMO water-cooled blanket based on supercritical water condition," *Fusion Engineering and Design*, *86*, Issue 12, 2839-2842.

[45] Boccaccini, L. V., Fischer, U., Gordeev, S. & Malang, S. (2000). "Advanced helium cooled pebble bed blanket with SiCf/SiC as structural material," *Fusion Engineering and Design*, *49–50*, 491–497.

[46] Raffray, R., Malang, S., El-Guebaly, L. & Wang, X. (2005). "Ceramic breeder blanket for ARIES-CS," *Fusion Science and Technology*, *47*, No. 3, 422-431.

[47] Nishio, S. et al., (1998). "Improved tokamak concept focusing on easy maintenance," *Fusion Engineering and Design*, *41*, 357–364.

[48] The ITER project: http://www.iter.org/.

[49] CANDU reactors: http://www.candu.org/candu_reactors.html.

[50] DOE Inspector General Report DOE IG-0632, December 2003; www.ig.energy.gov/documents/CalenderYear2003/ig-0632.pdf.

[51] El-Guebaly, L. & Malang, S. (2009). "Toward the ultimate goal of tritium self-sufficiency: technical issues and requirements imposed on ARIES advanced fusion power plants," *Fusion Engineering and Design*, *84*, 2072-2083.

[52] El-Guebaly, L., Jaber, A. & Malang, S. (2012). "State-of-the-art 3-D assessment of elements degrading the TBR of the ARIES DCLL blanket," *Fusion Science and Technology*, *61*, #4, 321-331.

[53] Batistoni, P., Angelone, M., Fischer, U., Klix, A., Kodeli, I., Leichtle, D., Pillon, M., Pohorecki, W. & Villari, R. (2012). "Tritium breeding experiments on HCPB and HCLL blanket mock-ups irradiated with 14 MeV neutrons," *Nuclear Fusion*, *52*, 083014.

[54] (2008). "DAGMC Users Guide," University of Wisconsin-Madison Fusion Technology Institute; https://trac.cae.wisc.edu/trac/svalinn/wiki/DAGMCUsersGuide.

[55] El-Guebaly, L. A. & Malang, S. (2009)."Need for online adjustment of tritium bred in blanket and implications for ARIES power plants," University of Wisconsin Fusion Technology Institute Report, UWFDM-1372. Available at: http://fti.neep.wisc.edu/pdf/fdm1372.pdf.

[56] Morgan, L. & Pasley, J. (2013). "Tritium breeding control within liquid metal blankets," *Fusion Engineering and Design*, *88*, 107-112.

[57] The Frascati Neutron Generator at ENEA, Italy: http://www.fusione.enea.it/LABS/FNG/detail_en.html.

[58] The Fusion Neuron Source at JAEA, Japan: http://fnshp.tokai-sc.jaea.go.jp/english/index_e.html.

[59] The TUD (Technische Universität Dresden) neutron generator at Dresden, Germany: http://www.fzd.de/db/Cms?pNid=321.

[60] Giancarli, L., et al., (2012). "Overview of ITER TBM program," *Fusion Engineering and Design*, *87*, 395-402.

[61] Zinkle, S. J., Busby, J. T. (2009). "Structural materials for fission and fusion energy," *Materials Today*, *12*, 12-19.

[62] Zinkle, S. J. & Möslang, A. (2013). "Evaluation of irradiation facility options for fusion materials research and development," *Fusion Engineering and Design*, *88*, 472-482.

[63] Sawan, M. E. (2012). "Damage parameters of structural materials in fusion environment compared to fission reactor irradiation," *Fusion Engineering and Design*, *87*, 551-555.

[64] Anderoglu, O., Byun, T. S., Toloczko, M. & Maloy, S. A. (2013). "Mechanical performance of ferritic martensitic steels for high dose applications in advanced nuclear reactors," *Metallurgical and Materials Transactions A*, *44*, S70-S83.

[65] Zinkle, S. J. & Ghoniem, N. M. (2000). "Operating temperature windows for fusion reactor structural materials," *Fusion Engineering and Design*, *51–52*, 55–71.

[66] Zinkle, S. J. & Snead, L. L. (2014). "Designing radiation resistance in materials for fusion energy," *Annual Review of Materials Research*, *44*, 241-267.

[67] Odette, G. R., Alinger, M. J. & Wirth, B. D. (2008). "Recent developments in irradiation-resistant steels," *Annual Review of Materials Research*, *38*, 471-503.

[68] Dai, Y., Henry, J., Tong, Z., Averty, X., Malaplate, J. & Long, B. (2011). "Neutron/proton irradiation and He effects on the microstructure and mechanical properties of ferritic/martensitic steels T91 and EM10," *Journal of Nuclear Materials*, *415*, 306–310.

[69] Tanigawa, H., Shiba, K., Möslang, A., Stoller, R. E., Lindau, R., Sokolov, M. A., Odette, G. R., Kurtz, R. J. & Jitsukawa, S. (2011). "Status and key issues of reduced activation ferritic/martensitic steels as the structural material for a DEMO blanket," *Journal of Nuclear Materials*, *417*, 9-15.

[70] Rieth, M., Schirra, M., Falkenstein, A., Graf, P., Heger, S., Kempe, H., Lindau, R. & Zimmermann, H. (2003). "Eurofer97: tensile, charpy, creep, and structural tests," FZKA-6911, *Forschungszentrum Karlsruhe*.

[71] Zinkle, S. J., Möslang, A., Muroga, T. & Tanigawa, H. (2013). "Multimodal options for materials research to advance the basis for fusion energy in the ITER era," *Nuclear Fusion*, *53*, 104024.

[72] Kurtz, R. J., Alamo, A., Lucon, E., Huang, Q., Jitsukawa, S., Kimura, A., Klueh, R. L., Odette, G. R., Petersen, C., Sokolov, M. A., Spätig, P. & Rensman, J-W. (2009). "Recent progress toward development of reduced activation ferritic/martensitic steels for fusion structural applications," *Journal of Nuclear Materials*, *386–388*, 411–417.

[73] Huang, Q., Baluc, N., Dai, Y., Jitsukawa, S., Kimura, A., Konys, J., Kurtz, R. J., Lindau, R., Muroga, T., Odette, G. R., Raj, B., Stoller, R. E., Tan, L., Tanigawa, H., Tavassoli, A-A.F., Yamamoto, T., Wan, F. & Wu, Y. (2013). "Recent progress of R&D activities on reduced activation ferritic/martensitic steels," *Journal of Nuclear Materials*, *442*, S2-S8.

[74] Raj, B., Rao, K. B. S. & Bhaduri, A. K. (2010). "Progress in the development of reduced activation ferritic-martensitic steels and fabrication technologies in India," *Fusion Engineering and Design*, *85*, 1460-1468.

[75] Odette, G. R. (2014). "Recent progress in developing and qualifying nanostructured ferritic alloys for advanced fission and fusion applications," *JOM*, *66*, 2427-2441.

[76] Wilshire, B. & Lieu, T. D. (2004). "Deformation and damage processes during creep of Incoloy MA957," *Material Science and Engineering A*, *386*, 81-90.

[77] Alamo, A., Lambard, V., Averty, X. & Mathon, M. H. (2004). "Assessment of ODS-14%Cr ferritic alloy for high temperature applications," *Journal of Nuclear Materials*, *329-333*, 333-337.

[78] Steckmeyer, A., Rodrigo, V. H., Gentzbittel, J. M., Rabeau, V. & Fournier, B. (2012). "Tensile anisotropy and creep properties of a Fe–14CrWTi ODS ferritic steel," *Journal of Nuclear Materials*, *426*, 182-188.

[79] Mukhopadhyay, D. K., Froes, F. H. & Gelles, D. S. (1998). "Development of oxide dispersion strengthened ferritic steels for fusion," *Journal of Nuclear Materials*, *258-263*, 1209-1215.
[80] Odette, G. R. & Hoelzer, D. T. (2010). "Irradiation-tolerant nanostructured ferritic alloys: transforming helium from a liability to an asset," *JOM*, *62*, 84-92.
[81] Gaganidze, E., Rupp, D. & Aktaa, J. (2014). "Fracture behaviour of polycrystalline tungsten," *Journal of Nuclear Materials*, *446*, 240–245.
[82] Hasegawa, A., Fukuda, M., Tanno, T. & Nogami, S. (2013). "Neutron irradiation behavior of tungsten," *Materials Transactions*, *54*, 466-471.
[83] Greenwood, L. R. & Garner, F.A. (1994). "Transmutation of MO, Re, W, Hf, and V in various irradiation test facilities and STARFIRE," *Journal of Nuclear Materials*, *212-215*, 635-639.
[84] El-Guebaly, L., Kurtz, R., Rieth, M., Kurishita, H. & Robinson, A. (2011). "W-based alloys for advanced divertor designs: options and environmental impact of state-of-the-art alloys," *Fusion Science and Technology*, *60*, Number 1, 185-189.
[85] Brooks, J. N., El-Guebaly, L., Hassanein, A. & Sizyuk, T. (2015). "Plasma facing material alternatives to tungsten," *Nuclear Fusion, 55*, 043002.
[86] Garin, P., Diegele, E., Heidinger, R., Ibarra, A., Jitsukawa, S., Kimura, H., Möslang, A., Muroga, T., Nishitani, T., Poitevin, Y., Sugimoto, M. & Zmitko, M. (2011). "IFMIF specifications from the users point of view," *Fusion Engineering and Design*, *86*, 611–614.
[87] ITER Specialized Tools: http://www.iter.org/construction/ assembly tooling
[88] Raffray, A. R. "Fusion technology information from ITER in-vessel components applicable to DEMO and beyond," to be published in Fusion Science and Technology.
[89] Badger, B., et al., (1974). UWFCM-68 "UWMAK-I, A Wisconsin toroidal fusion reactor," UWFCM-68. Available at: http://fti.neep.wisc.edu/pdf/fdm68.pdf.
[90] Waganer, L., Peipert, R., Wang, X., Malang, S. & the ARIES Team. (2008). "ARIES-CS maintenance system definition and analysis," *Fusion Science and Technology*, *54*, No. 3, 787-817.
[91] Wang, X. R., Tillack, M. S., Najmabadi, F., Malang, S. & the ARIES Team. (1999). "Configuration and Maintenance of the ARIES-ST Power Plant," 18th IEEE Symposium on Fusion Engineering, Oct. 24-28, Albuquerque, NM.

[92] The ESBWR Design: http://www.gepower.com/prod_serv/products/nuclear_energy/en/new_reactors/esbwr.htm.

[93] El-Guebaly, L., Wilson, P., Henderson, D. & Varuttamaseni, A. (2004). "Feasibility of target materials recycling as waste management alternative," *Fusion Science and Technology*, *46*, 506-518.

[94] Zucchetti, M., El-Guebaly, L., Forrest, R., Marshall, T., Taylor, N. & Tobita, K. (2007). "The feasibility of recycling and clearance of active materials from a fusion power plant," *Journal of Nuclear Materials*, *367-370*, 1355-1360.

[95] Anigstein, R., et al., (2003). "Radiological assessments for clearance of materials from nuclear facilities," volume 1, NUREG-1640, US Nuclear Regulatory Commission, Available at: http://www.nrc.gov/ reading-rm/doc-collections/nuregs/staff/sr1640/.

[96] (2004). International Atomic Energy Agency, Application of the concepts of exclusion, exemption and clearance, IAEA Safety Standards Series, No. RS-G-1.7. Available at: http://www-pub.iaea.org/MTCD/publications/PDF/Pub1202_web.pdf.

[97] El-Guebaly, L. (2007). "Evaluation of disposal, recycling, and clearance scenarios for managing ARIES radwaste after plant decommissioning," *Nuclear Fusion*, *47*, 485-488.

[98] El-Guebaly, L., Massaut, V., Tobita, K. & Cadwallader, L. (2008). "Goals, challenges, and successes of managing fusion active materials," *Fusion Engineering and Design*, *83*, 928-935.

[99] Massaut, V., Bestwick, R., Brodén, K., Di Pace, L., Ooms, L., & Pampin, R. (2007). "State of the art of fusion material recycling and remaining issues," *Fusion Engineering and Design*, *82*, 2844-2849.

[100] Zucchetti, M., Di Pace, L., El-Guebaly, L., Kolbasov, B. N., Massaut, V., Pampin, R. & Wilson, P. (2009). "The back end of the fusion materials cycle," *Fusion Science and Technology*, *52*, 109-139.

[101] Di Pace, L., El-Guebaly, L., Kolbasov, B., Massaut, V. & Zucchetti, M. (2012). "Radioactive Waste Management of Fusion Power Plants," Chapter 14 in Book: Radioactive Waste. Dr. Rehab Abdel Rahman (Ed.), ISBN: 978-953-51-0551-0, InTech. Available at http://www.intechopen.com/books/radioactive-waste/radioactive-waste-management-of-fusion-power-plants.

[102] Kolbasov, B., El-Guebaly, L., Khripunov, V., Someya, Y., Tobita, K. & Zucchetti, M. (2014). "Some technological problems of fusion materials management," *Fusion Engineering and Design*, *89*, 2013-2017.

[103] El-Guebaly, L. A. & Cadwallader, L. (2014). "Perspectives of managing fusion radioactive materials: technical challenges, environmental impact, and US policy." Chapter in book: Radioactive Waste: Sources, Management and Health Risks. Susanna Fenton Editor. NOVA Science Publishers, Inc.: Hauppauge, New York, USA. ISBN: 978-1-63321-731-7.

[104] (2014). US Code of Federal Regulations, Title 10, Energy, Part 61, Licensing Requirements for Land Disposal of Radioactive Waste, Section 55, Waste classification, US Government Printing Office.

[105] Fetter, S., Cheng, E. T. & Mann, F. M. (1990). "Long term radioactive waste from fusion reactors: part II," *Fusion Engineering and Design*, *13*, 239-246.

[106] El-Guebaly, L., Wilson, P. & Paige, D. (2006). "Evolution of clearance standards and implications for radwaste management of fusion power plants," *Fusion Science and Technology*, *49*, 62-73.

In: Fusion Energy and Power
Editor: Lionel Romero
ISBN: 978-1-63482-548-1

Chapter 2

OVERVIEW OF FUSION NEUTRONICS EXPERIMENTS FOR BLANKET OF A HYBRID ENERGY REACTOR

Rong Liu*

Institute of Nuclear Physics and Chemistry, Key Laboratory of Neutron Physics, China Academy of Engineering Physics, Mianyang, China

ABSTRACT

Advances in fusion neutronics integral experiments in the benchmark mock assemblies for the blanket of a fusion-fission hybrid energy reactor is described in brief. The concept of the hybrid energy reactor is that fusion energy from magnetic confinement fusion device is multiplied by released fission energy from the subcritical reactor. A conceptual blanket with the subcritical reactor is mainly loaded with natural uranium and lithium material. In the fission fuel region, uranium material and light water are arranged alternately. The integral experiments are performed for validating nuclear data and code in conceptual design. Based on materials available, the mock assemblies of the blanket are established. The spherical fission assemblies for fission research and plutonium production consist of three layers of depleted uranium shells and several layers of polyethylene shells. The spherical lithium assembly for tritium production consists of depleted uranium and LiPb alloy shells. The cubic fission assembly consists of natural uranium and polyethylene and its

* Author: liurongzy@163.co.

structure is basically consistent with one of the fission fuel region. In the assemblies with the D-T fusion neutron source, the plutonium production rates, uranium fission rates, ratios of capture to fission, ^{238}U (n, 2n) reaction rates and tritium production rates are measured, separately. The measured results are compared to the calculated ones with MCNP code and ENDF/B-VI library data.

Keywords: fusion-fission hybrid energy reactor, mockup, integral experiment, reaction rate, Monte Carlo simulation

1. INTRODUCTION

The fusion-fission hybrid energy reactor (FFHER) is designed in China Academy of Engineering Physics and consists of magnetic confinement fusion device and a subcritical reactor [1-2]. FFHER's concept is that fusion energy is multiplied by released fission energy from the subcritical reactor. The uranium fuel of the subcritical reactor as a fission blanket can be natural uranium (NU), LWR spent fuel, or depleted uranium. The design can increase the utilization rate of uranium. It would be one of the effective ways of applying fusion technologies to solve the future energy crisis. A conceptual blanket of the hybrid reactor is mainly loaded with uranium (UZr alloy) as nuclear fuel and lithium material (Li_4SiO_4) as tritium breeder. The schematic diagram of the conceptual blanket is shown in Figure 1. In the fission fuel region, the uranium and light water are arranged alternately. Light water is coolant and moderator. The structure of the fuel region is lattice cell. The volume ratio of NU to light water is about 2:1. In the uranium, fertile nucleus ^{238}U is used to breed ^{239}Pu by ^{238}U capture reaction, and fission nuclei are used to multiply fusion energy by fission reaction. In the tritium production region, fertile nuclide ^{6}Li is used to breed T for maintaining tritium self-restraint by ^{6}Li (n,α) T reaction. The hybrid energy reactor has been never operated. In order to deal with various nuclear problems originating from D-T neutrons in applications for hybrid energy, it is crucial to validate nuclear data and code used in the design, and check neutron property of the conceptual blanket by neutronics integral experiments.

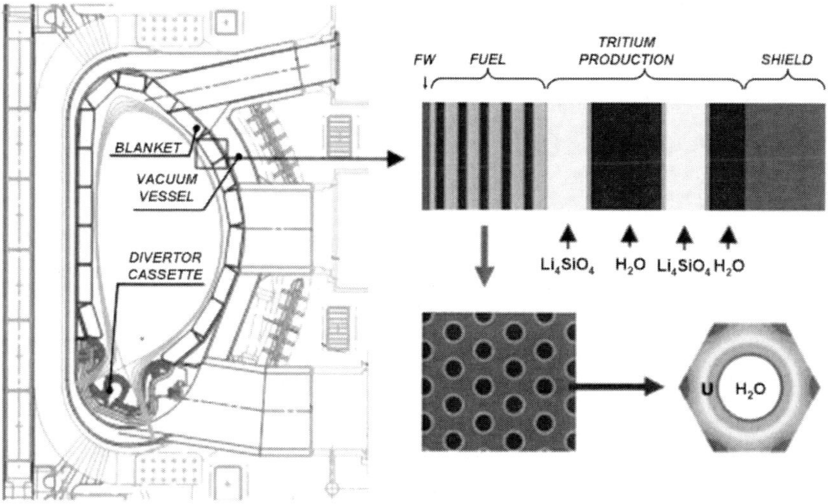

Figure 1. Schematic diagram of the conceptual blanket.

We have performed a series of benchmark integral experiments for validating nuclear data and code used in the design in Institute of Nuclear Physics and Chemistry (INPC) [3-8]. In the first stage experiments, spherical shells available were combined to setup benchmark mock assemblies in which neutron spectra similar to ones in the conceptual blanket. The spherical assemblies are different combinations of depleted uranium, polyethylene or LiH spherical shells. Depleted uranium is used for fission fuel, while polyethylene for light water and LiH for tritium production. The reaction rates and leakage neutron energy spectra were measured in the assemblies. The uncertainty of reaction rates is analyzed. The experimental results were calculated by the MCNP code employing ENDF/B-VI library data, and the ratios of calculation to experiment (C/E) were obtained. The results indicated that the code and relevant nuclear data were applicable to analyze the integral experiments. It was concluded that a special database needs to be established and validated before the physics design of the conceptual blanket, in particular a database for the thermal neutron to ensure the feasibility of the design. In addition, the neutronics integral experiments relevant to fusion neutrons in other laboratories were performed [9-10], and the difference of ^{238}U capture reaction rates between experiment and calculation was generally about 10%.

Based on demand for checking neutron property of the conceptual blanket, we have implemented the integral experiments further in the second stage [11].

Referred to the blanket design, we have established the analytical mockups which are integral parts of the whole blanket. One kind of mockups is still spherical and mainly consisted of depleted uranium and polyethylene shells, available. In the spherical mockups, graphite reflector and iron shells for simulating the first wall are increased, and LiPb alloy shell is used to simulate a breeder. Another kind of mockup is cubic and its structure and material are basically consistent with ones of the fission fuel region. The integral experiments in the mockups have been performed. The reaction rates are measured in the mockups.

In this chapter, main advances in fusion neutronics integral experiments in the mock assemblies (as mockups) for the conceptual blanket of FFHER in INPC are described in brief. Representative results of integral experiments with D-T fusion neutrons are collected and presented. The mock assemblies include the spherical fission assembly with depleted uranium/polyethylene shells and the cubic assembly with natural uranium/polyethylene materials. In the assemblies with the D-T fusion neutron source, the plutonium production rates, uranium fission rates, ratios of capture to fission, $^{238}U(n,2n)$ reaction rates and tritium production rates are measured, separately. The measured results are compared to the calculated ones with MCNP code and ENDF/B-VI library data.

2. Mockups

Aiming at validating nuclear data and code, and researching the neutron property of a hybrid reactor blanket, the ways for establishing mockups are that the neutron spectra, material and structure are relevant to the blanket.

2.1. Spherical Fission Mockup with DU/PE

The schematic diagram of spherical fission mockup with DU/PE is shown in Figure 2. This mockup consists of depleted uranium (~99.6% ^{238}U, ~0.4% ^{235}U) available and polyethylene, and the neutron spectra in the mockups are near to ones in the blanket. Polyethylene has similar property to light water as a neutron moderator. The inner radius (IR) and the outer radius (OR) of the assembly are 131 mm and 350 mm, respectively. Each polyethylene layer is sandwiched between two layers of DU. The thicknesses of three layers of DU, from inner to outer, are 50, 39 and 46 mm, respectively. The thicknesses of the three layers of polyethylene, from inner to outer, are 13, 21 and 50 mm, respectively. The outer polyethylene shell is also used to shield background neutrons increase counts at far positions to core. The volume ratio of depleted uranium (DU) to polyethylene (PE) is about 0.94:1, as the second assembly in ref. [8]. Moreover, in ref. [8], the first assembly with the volume ratio of about 2:1 didn't have the outer polyethylene shell and the inner polyethylene shell, while the third assembly with the volume ratio of DU to polyethylene of about 0.91:1 was constructed by adding inner polyethylene shell with 20 mm in thickness. The abundance of uranium in the assemblies is the same as ones in detectors of the DU foils and fission chamber [3].

A D-T fusion neutron source is placed at the center of the spherical mockup. In the 14-MeV neutron generator, the energy of D^+ ions is 225 keV and neutrons are produced by D^+ beam bombarding a TiT Target. The yield is about 3×10^{10} neutrons/s. The absolute yield of the neutron source is measured by the associated α particle method. An Au-Si surface barrier semiconductor detector is at 178.2^0 to incident D^+ beam direction in drift tube and used to measure the yield.

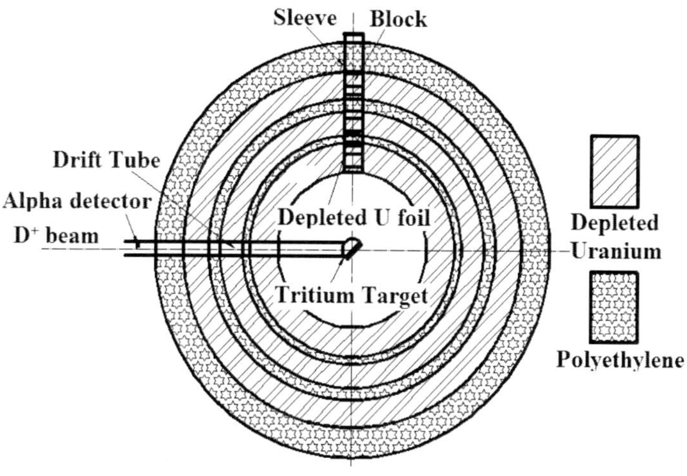

Figure 2. Spherical fission mockup with alternate DU/PE shells.

2.2. Spherical Fission Mockup with Fe/DU/PE/GRAPHITE

The spherical fission mockup with Fe/DU/PE/Graphite shown in Figure 3 consists of DU, PE, graphite and iron. The graphite shell is used as reflector and iron shells for mimicking the first wall. IR and OR of the mockup are 111 mm and 365 mm, respectively. Its volume ratio of DU to PE is about 0.94:1. Each polyethylene layer is sandwiched between two layers of DU. The thickness of three layers of DU, from inner to outer, are 50, 39 and 46 mm, respectively. The thickness of the two layers of polyethylene, from inner to outer, are 13 and 21 mm, respectively. The outer graphite shell of 300-mm IR and 350-mm OR and iron shell of 350-mm IR and 365-mm OR are also used to shield background neutrons and increase counts at far positions to core. The inner iron shell is 111 mm in IR and 131 mm in OR. A D-T fusion neutron source is placed at the center of the spherical mockup.

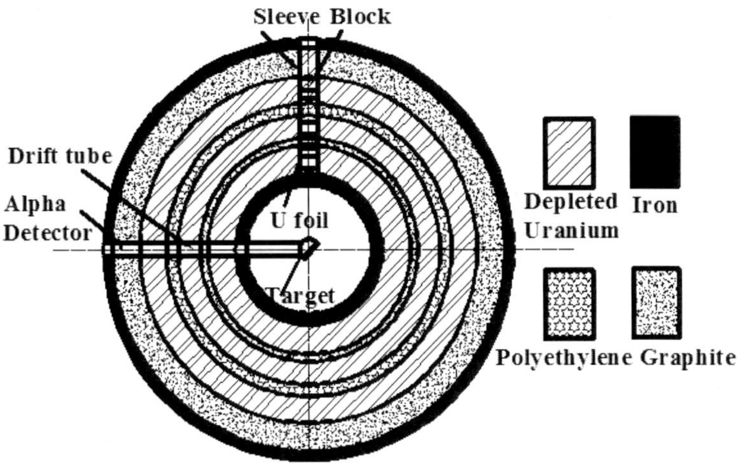

Figure 3. Spherical fission mockup with Fe/DU/PE/Graphite shells.

2.3. Spherical Lithium Mockup

The schematic diagram of the spherical lithium mockup for tritium production to mimic the blanket is shown in Figure 4 [11]. It is of 40-mm IR and 254-mm OR, with a DU shell of 91 mm thicknesses, a LiPb alloy shell of 102 mm thickness and a PE shell of 21 mm thickness. The DU shell is used to mimic fission fuel region so that neutrons with fission spectrum can leak into the LiPb region for tritium production. The portion of ^{nat}Li in LiPb is 18%. The outer PE shell of 233-mm IR and 254-mm OR is used to shield background neutrons and increase counts at far positions to core. A D-T fusion neutron source is placed at the center of the spherical mockup.

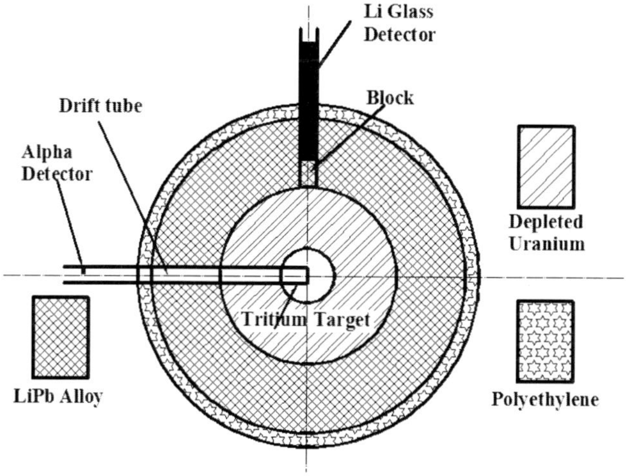

Figure 4. Spherical lithium mockup with depleted-uranium and LiPb shells.

2.4. Cubic Fission Mockup

The schematic diagram of the cubic fission assembly to mimic the fission blanket is shown in Figure 5 [11]. This mockup with lattice cell structure consists of NU (~99.3% ^{238}U, ~0.7% ^{235}U) and PE, and its material and structure are basically consistent with ones of the fission blanket. The length and height of the mockup are 467.5 mm and 460 mm, the thickness 206 mm, respectively. The 195 holes of 8.5-mm radius are equably distributed in the NU mockup. The PE tubes of 6-mm IR and 8.5-mm OR have the shape of hollow cylinder and are placed in the holes. Its volume ratio of NU to PE is about 2.3:1. The PE is used to mimic light water as a coolant and neutron moderator as described above.

A D-T fusion neutron source is placed at normal distance of 100 mm to the center of the cubic mockup surface. The detector is placed in holes of 230-mm height to the central level surface of mockup. The abundance of uranium in the mockup is the same as ones in detectors of the NU foils in ring shape.

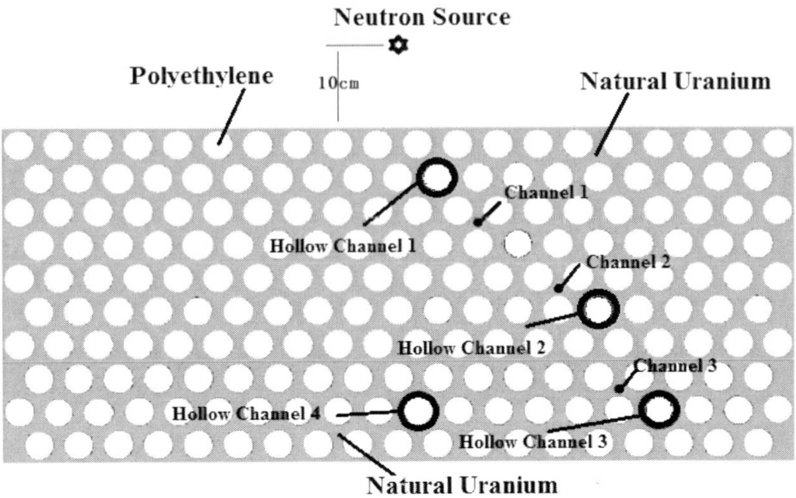

Figure 5. Cubic fission mockup with natural uranium and polyethylene.

3. Results in Spherical Fission Mockup

3.1. ^{238}U Capture Reaction Rates

The fertile nuclide ^{238}U capture reaction rate indicates breeding of fissile fuel ^{239}Pu. ^{238}U (n,γ) ^{239}U capture reaction rates by measuring 277.6 keV γ rays emitted from ^{239}Np (^{239}U decay) in DU foils, i.e., the production rate of ^{239}Pu (^{239}Np decay), can be deduced [3-4].

The 18 slices of DU foils with 24 mm in diameter and 0.122~0.193 mm in thickness are put in the radial channel of the DU/PE shells at a 90^0 direction to incident D^+ ion beam as shown in Figure 2. The 7 slices of DU foils with 24 mm in diameter and 0.179 mm ~0.396mm in thickness are put in the radial channel of the Fe/DU/PE/Graphite shells at a 90^0 direction to incident D^+ ion beam as shown in Figure 3.The self-absorption correction factors of a DU foil to 277.6 keV γ rays are measured with the HPGe gamma spectrometer and ^{243}Am γ ray source. The values obtained by the least square method are 93.3%~89.8% in DU/PE shells and 90.4%~80.4% in Fe/DU/PE/Graphite shells depending on thicknesses of DU foils. The source neutron yield during irradiation is measured by an auto-timing counts system and corrected by the decay factor [3-4].

The plutonium production rates (PPR) are obtained and normalized to one source neutron and one ^{238}U atom. The uncertainty of PPR is 3.7%. The sources of PPR uncertainty include the following respects: neutron source monitoring about 2.5%; γ detecting efficiency 1.9%; γ emission branch ratio 1.5%; self-absorption 1.0%; γ count statistics 0.5%~0.7%.

The PPR is calculated using MCNP code and ENDF/B-VI library data. The model is completely consistent with the structure and ingredient of the assembly contained the target chamber. The S (á, â) thermal scattering model in PE is considered. The angular dependences of the source neutron energy and intensity were calculated by "DROSG-2000" code [12]. The distribution of the source neutron energy with angle is described. The calculated error is less than 1%. The PPR distributions of experiment and calculation in DU/PE shells are shown in Figure 6 [8]. The ranges of C/E (ratio of Calculation to Experiment) values are 1.01~1.10 and the average C/E value 1.05.The PPR distributions of experiment and calculation in Fe/DU/PE/Graphite shells are shown in Figure 7 [11]. The range of C/E values is 1.04~1.13 and the average C/E value 1.09.

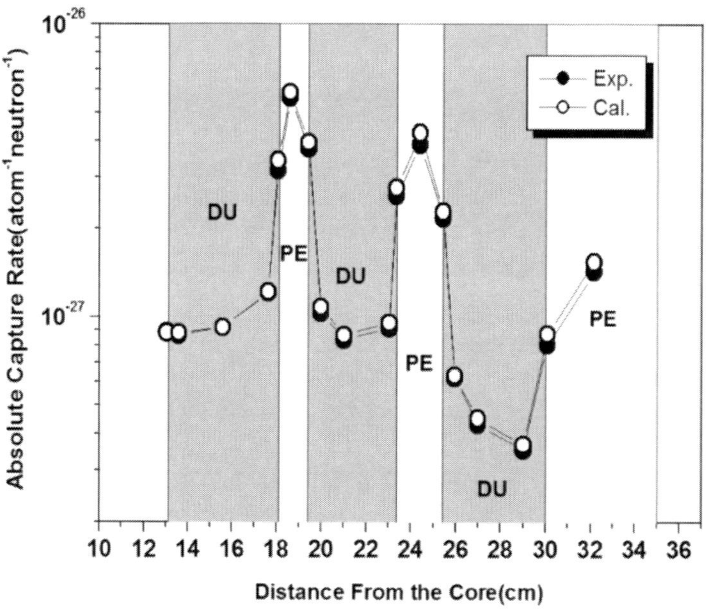

Figure 6. Distributions of ^{238}U capture reaction rates in DU/PE shells.

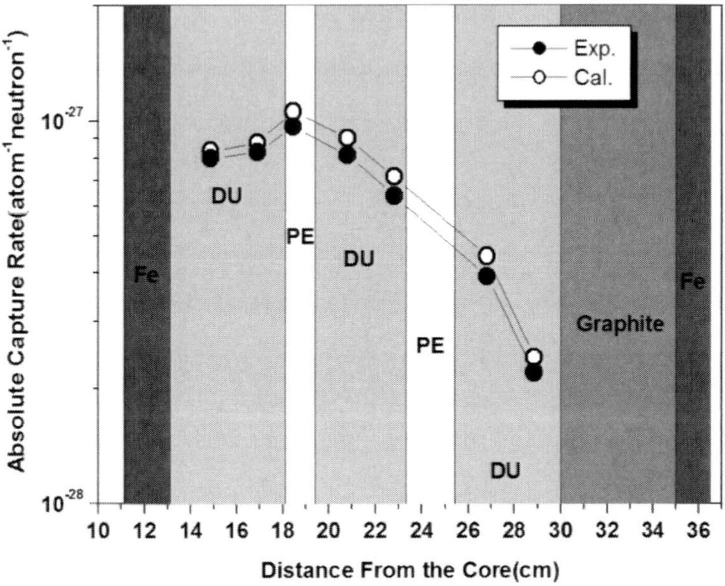

Figure 7. Distributions of ^{238}U capture reaction rates in Fe/DU/PE/Graphite shells.

3.2. Reaction Rate Ratio of Capture to Fission

The reaction rate ratio of ^{238}U capture to total fission gives fissile production rate in unit of fuel burn-up in the fission blanket. The ratio of ^{238}U capture to total fission can be obtained by using a DU foil put in uranium region of a mockup. While the ^{238}U capture reaction rate measured above, the total fission rate is simultaneously measured by using the activation method and the fission yield correction method [5]. 293.3keV γ rays from U fission product ^{143}Ce were measured for the total fission rate. The uncertainties of relative ratios are 3.7%~5.7%. The ratios are calculated by the MCNP code employing ENDF/B-VI library data.

The relative ratios are about 1~2 in DU regions of DU/PE shells and the range of C/E values is 0.93~1.06 shown in Figure 8 [5]. The relative ratios are about 1.39~2.04 in DU regions of Fe/DU/PE/Graphite shells and the range of C/E values is 0.90~1.08 shown in Figure 9. In addition, the ratios are about 0.91~1.54 in depleted uranium shell with 194 mm/300 mm in IR/OR without PE. It is indicated that adding PE in the spherical fission assembly can enhance fissile fuel production rate.

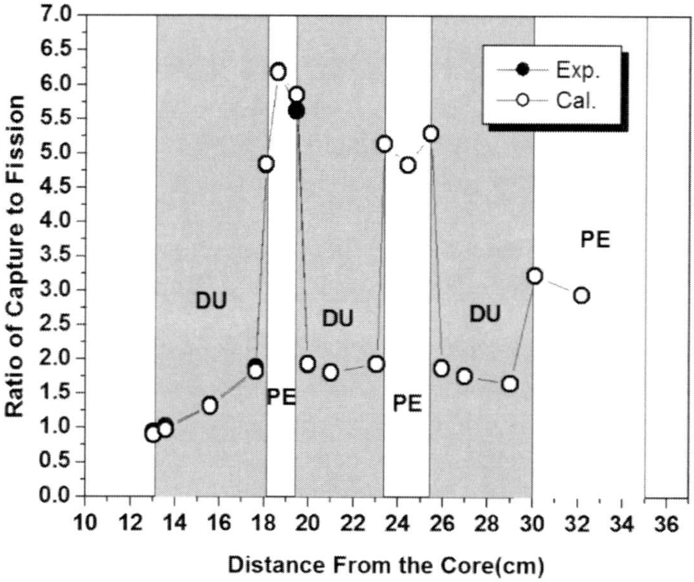

Figure 8. Ratio of capture to fission in DU/PE shells.

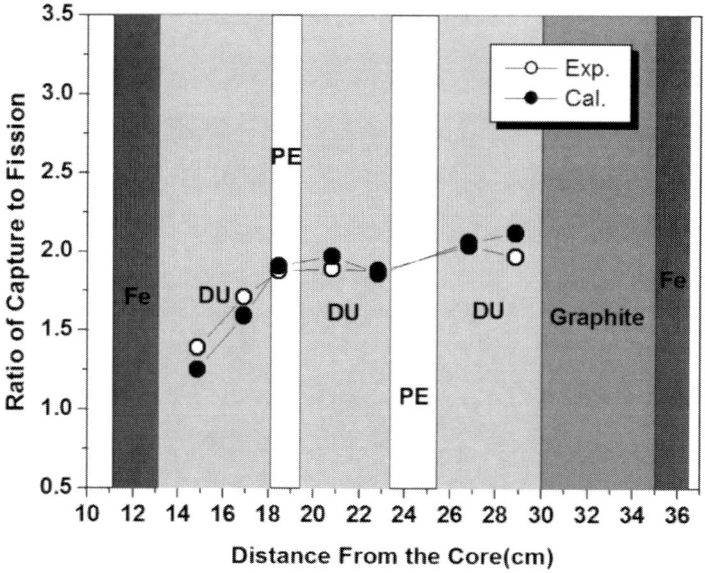

Figure 9. Ratio of capture to fission in Fe/DU/PE/Graphite shells.

3.3. Uranium Fission Rates

The fission reaction rate indicates the energy amplification and neutron multiplication. The plate DU fission chamber with 30 mm in diameter and 20 mm in height is used to measure fission rates [3, 6]. It is put in the radial channel, at a 90^0 direction to incident D^+ ion beam, of the spherical fission mockups. The DU fission rates (DUFR) are measured at 17 positions in the mockups. The DUFR is normalized to one source neutron and one DU atom in the fission chamber with the DU coating of 24 mm in diameter. The experimental uncertainty is about 4.0%.

The DUFR is calculated using MCNP code and ENDF/B-V library data. The model is completely consistent with the structure and ingredient of the assembly contained the target chamber. The calculated uncertainty is less than 2%. The DUFR distributions of experiment and calculation in DU/PE shells are shown in Figure 10 [8]. The ranges of C/E values is 1.05~1.17 in DU regions and the average C/E value 1.11 in DU regions. The DUFR distributions of experiment and calculation in Fe/DU/PE/Graphite shells are shown in Figure 11 [11]. The result states that the range of C/E values is 0.93~1.08 and the average C/E value 0.99 in DU regions.

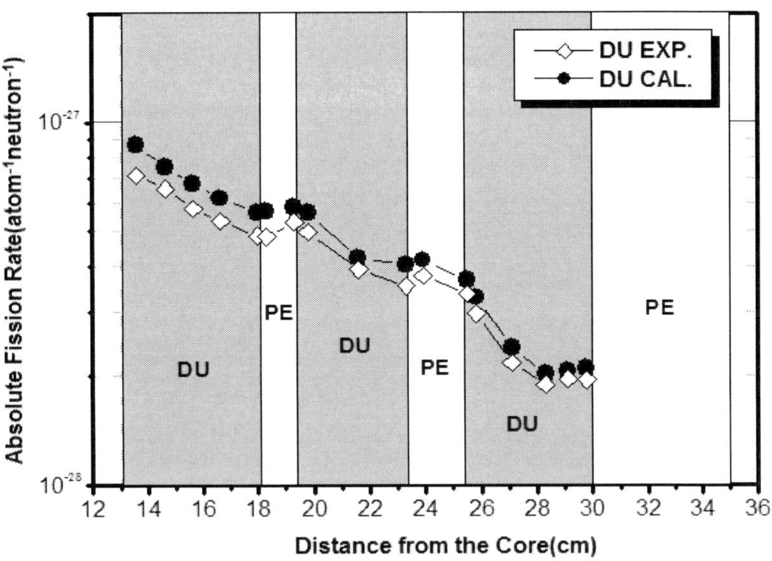

Figure 10. Distributions of depleted-uranium fission rates in DU/PE shells.

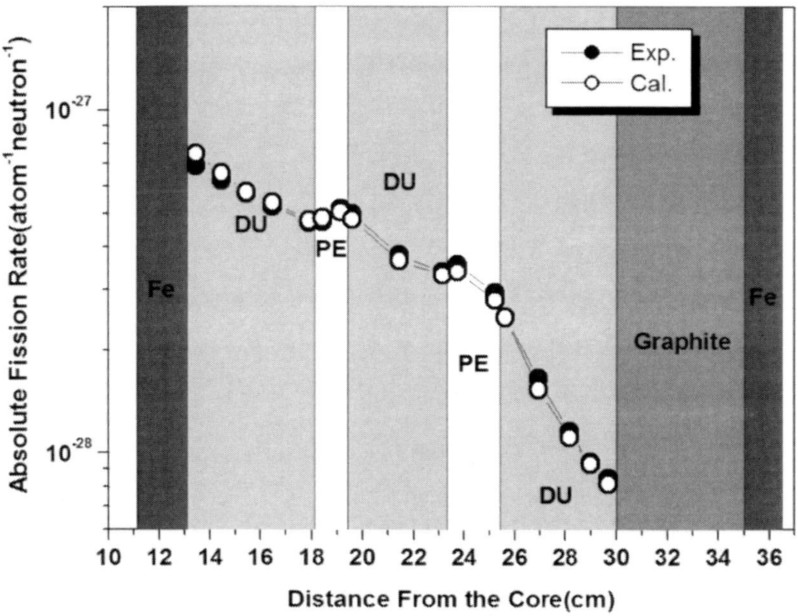

Figure 11. Distributions of depleted-uranium fission rates in Fe/DU/PE/Graphite shells.

3.4. ^{238}U(n,2n) Reaction Rates

In the fission blanket, the ^{238}U(n,2n) ^{237}U reaction with threshold energy of about 6 MeV is almost the only reaction for producing neutrons, except for fission. It is very significant to carry out ^{238}U(n,2n) reaction rate experiment for checking concept design of subcritical reactor. We carry out ^{238}U(n,2n) reaction rate experiment in the DU/PE shell as shown in Figure 2. ^{238}U(n,2n) ^{237}U reaction rates are deduced by measuring 208keV γ ray emitted from ^{237}U in DU foils. The 4 slices of DU foils with 24 mm in diameter and 0.15~0.19 mm in thickness are put in the radial channel of the 1st shell in DU/PE shells at a 90^{0} direction to incident D^{+} ion beam. The uncertainties of ^{238}U(n,2n) reaction rates are 5.3%~6.0%. The reaction rates are calculated using MCNP code and ENDF/B-VI library data. The calculated reaction rates are compared with the experiments in Figure 12 [6]. The difference between calculations and experiments are about 5%, and the average C/E value is 1.05.

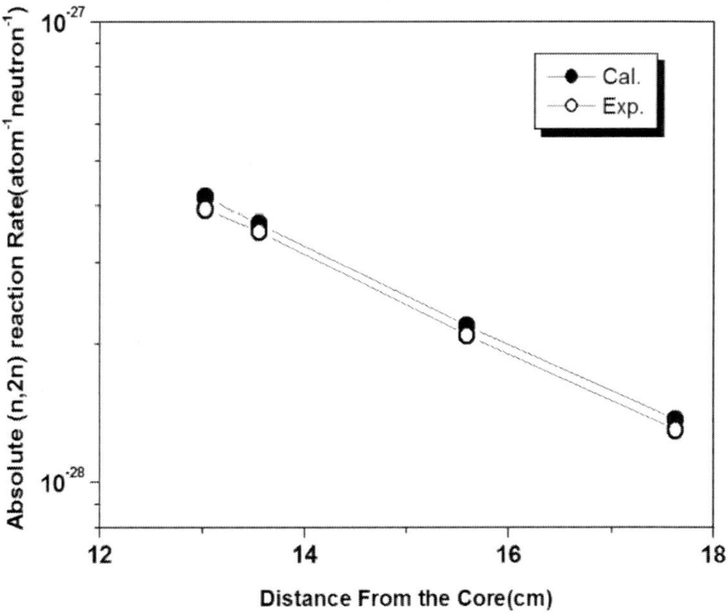

Figure 12. Comparison of measured and calculated ^{238}U(n,2n) reaction rate in DU/PE shells.

4. Tritium Production Rates in Spherical Lithium Mockup

The ^6Li (n,α) T reaction rate indicates the tritium production rate (TPR) in the spherical lithium mockup. The tritium production rate is measured by a ^6Li glass scintillation detector [3, 7]. The ^6Li glass detector with 2.62 mm in diameter and 0.22 mm in thickness is placed in the radial channel in lithium region of the mockup at 90° to D^+ beam incidence, separately, as shown in Figure 4. The reaction rates are normalized to one source neutron and one ^6Li atom in the ^6Li glass detector. The background counts from γ-rays in a ^6Li glass detector are subtracted by the counts from γ-rays in a ^7Li glass detector. The lithium glass detector is placed at 21 positions in 5-mm intervals. When the reaction rate is measured on-line at a position, a cavity in the front of the detector is filled using small LiPb cylinders with 25 mm in diameter and 5 mm in thickness.

The distribution of tritium production rates with an uncertainty of about 5.0% is normalized to one source neutron and shown in Figure 13 [11]. Because the leakage neutron spectrum from the DU shells is hard and neutrons are reflected by the PE shell, the reaction rates increased with increasing radial distance. The tritium production rates are calculated by MCNP code and ENDF/B-VI library data. The model is completely consistent with the structure and ingredient of the mockup contained the target chamber and the detector. The calculated uncertainty is less than 3%. The result states that the ranges of C/E values are 0.97~1.15 and the average C/E value 1.04. As for no effect of thermal neutron scattering of hydrogen in LiPb alloy, the calculated and experimental results are basically consistent within uncertainty.

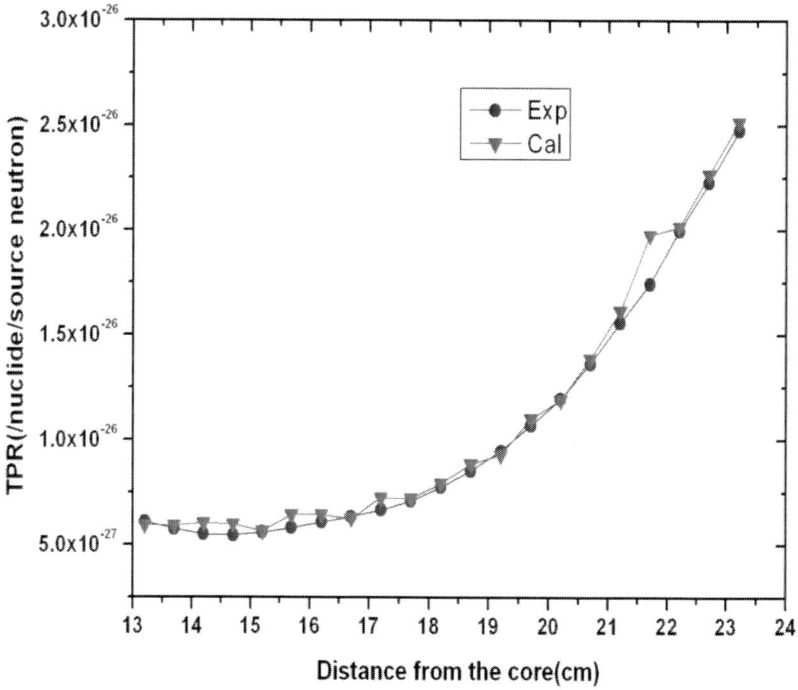

Figure 13. Distributions of tritium production rates in spherical lithium mockup.

5. Results In Cubic Fission Mockup

5.1. ^{238}U Capture Reaction Rates

The distributions of ^{238}U capture reaction rates, i.e., PPR, are measured at the four hollow channels with 8.5mm/12.15mm in IR/OR and 230mm in length as shown in Figure 5. The 9 slices of NU ring foils with 8.5mm/12.15mm in IR/OR and 0.34mm~0.46mm in thickness are put on a PE tube of 6mm/ 8.5mm in IR/OR. The NU hollow cylinder is filled between the two NU ring foils. The PE tube with NU ring foils and NU hollow cylinders is put into the hollow channels. The source neutron yield during irradiation is corrected by the decay factor [3]. The self-absorption correction factors of a NU foil to 277.6 keV γ rays are measured with the HPGe gamma spectrometer and ^{243}Am γ ray source, and the values are 76.9%-82.4% for NU foils with different thicknesses. The efficiency of the HPGe gamma spectromcter to 277.6 keV γ rays in the ring foil is calculated by using MCNP code and the value is 10.5%.

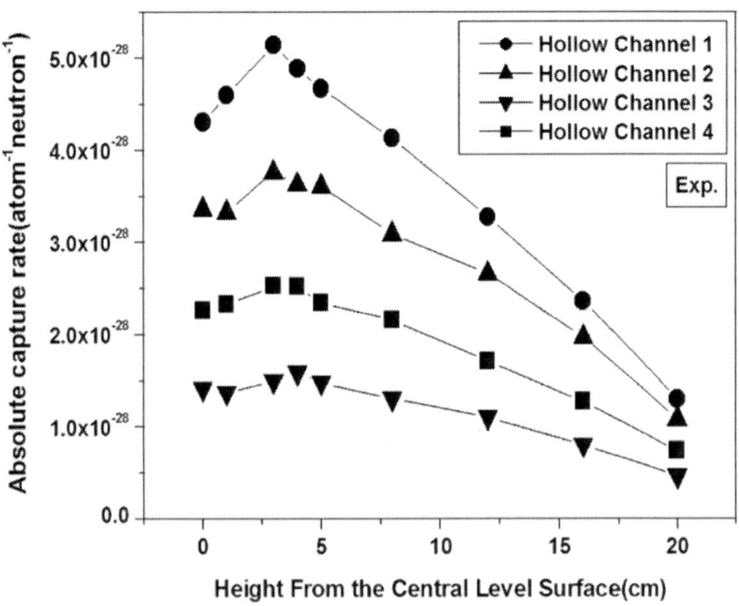

(a) Experimental results

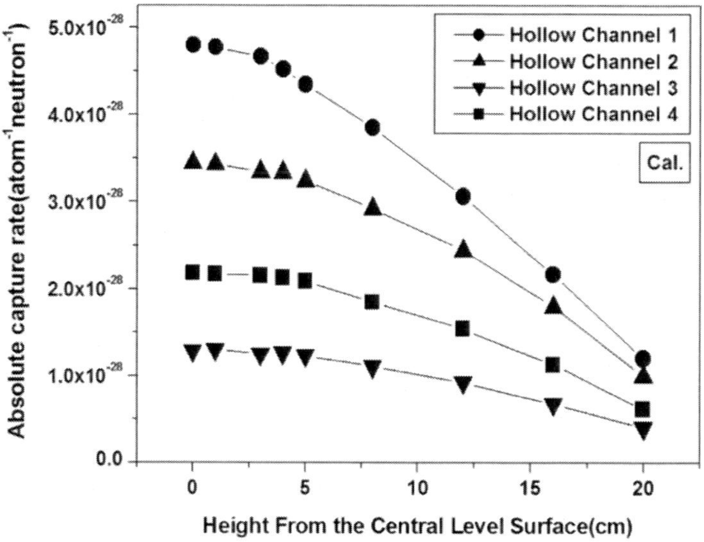

(b) Calculated results

Figure 14. Distributions of ^{238}U capture reaction rates in cubic fission mockup.

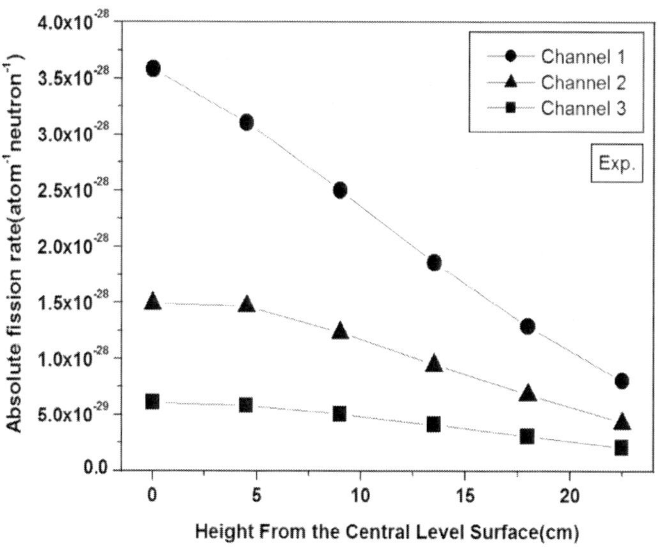

(a) Experimental results

Figure 15. (Continued).

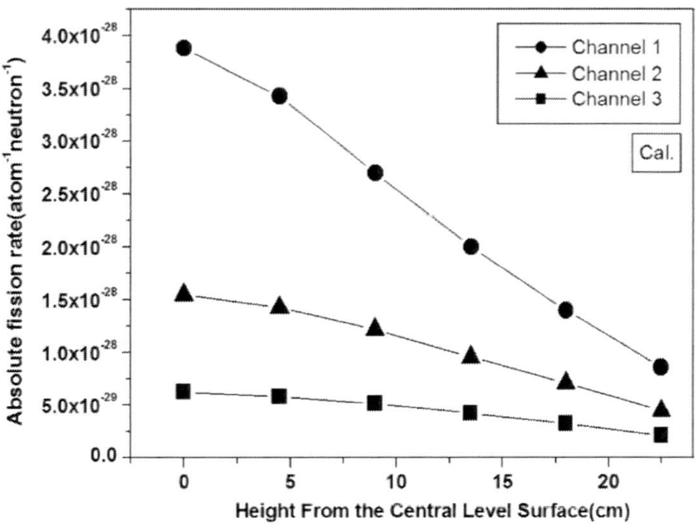

(b) Calculated results

Figure 15. Distributions of uranium fission rates in cubic fission mockup.

The preliminary PPR is obtained and normalized to one source neutron and one ^{238}U atom [11]. The uncertainties of PPR are 4.4%~4.5%. The experimental PPR distributions are shown in Figure 14 (a). The abscissa is vertical height from the central level surface of the mockup. The PPR is calculated using MCNP code and ENDF/B-VI library data. The model is completely consistent with the structure and ingredient of the mockup contained the target chamber. The calculated error is less than 2.4%. The range of C/E values is 0.84~1.11and the average C/E value 0.91. The calculated PPR distributions are shown in Figure 14 (b).

5.2. Uranium Fission Rates

The cylinder uranium fission chamber (99.997% ^{238}U, 0.003% ^{235}U) with 6.5mm in diameter and 25.4 mm in height is used to measure fission rates. It is put in the three vertical channels with 7mm in diameter separately, of the cubic mockup as shown in Figure 5. The uranium fission rates are measured at 6 positions in each channel and normalized to one source neutron and one uranium atom [11]. The experimental uncertainty is about 6.4%. The experimental distributions of uranium fission rates are shown in Figure 15 (a). The distributions were calculated using MCNP code and ENDF/B-V library data. The model is completely consistent with the structure and ingredient of the mockup contained the target chamber and the fission chamber. The calculated uncertainty is less than 1%. The result states that the range of C/E values is 0.97~1.11 and the average C/E value 1.04. The calculated distributions of uranium fission rates are shown in Figure 15 (b).

CONCLUSION

To validate nuclear data and code, and check the neutronics design of the conceptual blanket in the fusion-fission hybrid energy reactor, the fusion neutronics integral experiments in the mockups have been performed. The benchmark spherical mockups mainly based on DU shells available, and the cubic mockup based on NU material are established. The distributions of plutonium production rates, uranium fission rates, ratios of capture to fission, ^{238}U(n,2n) reaction rates and tritium production rates in the mockups are obtained, respectively. The experimental results in series of mock assemblies are self-consistent. The distributions are calculated with MCNP code and ENDF/B-VI library data. The C/E values are obtained. It is indicated that calculations in mockups generally agree with experiments within 10%. It is concluded that the code and library data used are adapted to the conceptual blanket design with D-T neutrons. The results are beneficial to perform the integral experiments in the test blanket module established by using the NU cube with PE tube and the tritium production region with Li_4SiO_4 further. The fusion neutronics integral experiments support the feasibility of the proposed physical design for the blanket in FFHER concept.

ACKNOWLEDGMENTS

This work is supported by the National Special Magnetic Confinement Fusion Energy Research, China, under contract No.2010GB111002 and No.2015GB108001, and the National Natural Science Foundation of China (No.91226104). The author wish to acknowledge the all participators of the projects, including Dr. Tonghua Zhu, Prof. Xinxin Lu, Prof. Xinhua Wang, Dr. Xiaosong Yan, Dr. Yiwei Yang, Dr. Song Feng, Prof. Li Jiang, Prof. Mei

Wang, et.al. The author would like to thank Prof. Benchao Lou and his group for their neutron generator operation.

REFERENCES

[1] Shi X.M.; Peng X.J. *Nucl Power Eng*. 2010, 314, 5-8.
[2] Li M.S.; Liu R.; Shi X.M. ; et al., *Fusion Eng Design*. 2012, 87, 1420-1424.
[3] Liu R.; Zhu T.H.; Yan X.S.; et al., *Nucl Sci Tech*. 2012, 23, 242-246.
[4] Yan X.S.; Liu R.; Lu X.X.; et al., *Acta Physica Sinica*. 2012, 61, 102801.
[5] Yang Y.W.; Yan X.S; Liu R.; et al., *Fusion Eng Design,* 2012, 87, 1679-1683.
[6] Yan X.S.; Liu R.; Lu X.X.; et al., *Chinese Physics C*. 2012, 36, 670-674.
[7] Wang X.H ; Guo H.P; Mou Y.F.; et al., *Chinese Physics C*. 2013, 37, 056202.
[8] Liu R. ; Zhu T.H; Yan X.S.; et al., *Nucl Data sheets,* 2014, 118, 588-591.
[9] Akiyama M.; Oka Y.; Kanasugi K.; et al., *Annals Nucl Energy,* 1987, 14, 543-553.
[10] Afana V.; Belevitin A.; Verzilov Y.; et al., *Atomnaya Energiya* , 1990, 71, 417-422.
[11] Liu R.; Zhu T.H.; Lu X.X.; et al., *Fusion Eng Design,* 2014, 89, 2994-2999
[12] Drosg M. IAEA-NDS-87, 2003, Rev. 8.

In: Fusion Energy and Power
Editor: Lionel Romero
ISBN: 978-1-63482-548-1

Chapter 3

OLD MATH AND RENEWED PHYSICS: KEYS TO ENGINEERING COLD FUSION

Cynthia Kolb Whitney
Galilean Electrodynamics

ABSTRACT

This chapter argues that Cold Fusion (CF) can become more understandable, and hence more amenable to Engineering, especially Control Engineering, if we will take account of appropriate redevelopments in Physics. These redevelopments become possible with fuller application of standard Mathematics. Of particular interest is the venerable protocol of formulating governing differential equations, identifying an appropriate family of solutions for the differential equations, and applying the boundary conditions that specify the particular problem at hand. This protocol leads to review of some results from Special Relativity Theory (SRT), which leads in turn to a better entrée into Quantum Mechanics (QM), and hence into Chemistry and Electro-chemistry, and finally into CF. We find the following practical recommendations re CF: 1) Let the heavy water rest quietly for a long time before attempting to elicit fusion events; 2) Consider both gravity and electrochemistry to anticipate how various ion species will distribute themselves in the water; 3) Study a range of possible arrangements of electrodes in the water to determine the optimum one.

INTRODUCTION

This Chapter builds upon a Chapter from my 2013 Nova book titled *Algebraic Chemistry: Applications and Origins.* [1] That Chapter was titled "Electro-Chemistry in Power Generation", and its subject was CF - the much-sought economically practical alternative to Hot Fusion (HF); i.e., thermonuclear fusion.

Fusion in general exploits the fact that Hydrogen atomic nuclei include not only Hydrogen H nuclei (one proton), but also Deuterium D nuclei (one proton and one neutron), and even Tritium T nuclei (one proton and two neutrons). In fusion, the neutrons are needed to create output Helium He nuclei.

HF exploits the phenomena of plasma state and magnetic confinement. In plasma state, nuclei are stripped 7naked of electrons. In magnetic confinement, the stripped nuclei can come together close enough that nuclear forces can overcome electrical repulsion and allow nuclear fusion, producing He nuclei, and releasing much energy.

CF exploits properties of water and electrochemistry. The heavy nuclei fuse at lower temperatures for reasons not yet completely understood. Viewed on the basis of old nuclei consumed, or new nuclei formed, CF can release equally as much energy as HF.

HF is easy to understand in terms of current-day Physics, but it is hard to produce, because a huge energy investment is required before any excess energy, i.e., energy beyond the input energy investment, can be harvested. By contrast, CF is difficult to understand in terms of current-day Physics, and without such understanding, its production remains mysterious and uncertain. As a result, we hardly even get to the issue of controlling it for commercial applications!

This Chapter argues that CF can become more understandable, and hence more controllable, if we will just review how theoretical developments in Physics happened historically, and then do the obviously indicated further developments.

Sometimes in reviewing history, the most intriguing events are the ones that inexplicably did *not* occur at the times when they clearly could have. In the case of electromagnetic signals as the foundation for much of twentieth century physics, several possible events did not actually happen:

1) In the early twentieth century, Albert Einstein was working at almost the same time on both the concept of the photon [2], which later led to

Quantum Mechanics (QM), and on the concept of signal, which was the basis for his development Special Relativity Thoery (SRT) [3]. But he did not connect the photon and the signal to each other.

2) In the middle of the twentieth century, Richard Feynman was developing Quantum Electrodynamics (QED) [4], which featured the concept of virtual photons as mediators of electromagnetic fields. But he did not then suggest reviewing SRT in light of the virtual-photon concept.
3) Throughout that same mid-century time frame, Claude Shannon was developing the founding concepts of present-day Information Theory (IT), conveniently collected in Brilloin's book [5]. IT was a powerful tool for wartime code breaking and code making, and for the post-war communication and computation industries. It really led to our current 'Age of Information'. But Shannon, did *not* suggest using IT to scrutinize the Signal concept at the foundation of SRT. Nor did anyone else at that time.

Why did none of these events happen? We do not know, and we cannot guess. It is, however, possible to remedy the situation now, and to do it in exactly the same way it could have been done so much earlier. It is all just Applied Mathematics.

In the late nineteenth century there was already a well-established protocol for dealing with problems involving differential equations. There are three steps:

1) Specify the differential equations;
2) Characterize the family of solutions that they allow;
3) Identify and use the boundary conditions that represent your particular problem.

Thus the recommended Applied Math approach consists of these tools: the governing differential equations (D), an appropriate family of solutions that satisfy the differential equations (S), and the boundary conditions (B) that specify the particular problem at hand. Together they make the triad DSB, and define the DSB Protocol for solving an Applied Math problem.

This Protocol is extremely well developed, widely used throughout Engineering Science and Applied Physics, and fully reliable. It should be applied to any problem that allows its application; it should *not* be skipped over arbitrarily.

Another aspect of recommended approach comes from even further back: the discipline instilled by the ancient Greek founder of Geometry, Euclid. [6] He began his development of Geometry with an absolutely minimal set of Axioms (noun of Greek origin) that are stated in the form "Let us Postulate... (verb of Latin origin; also used as a noun equivalent to 'Axiom' in modern texts). Axioms or Postulates are not to be multiplied beyond the minimum possible set; an extra Postulate can only be redundant, or worse: conflictive. And if conflictive, an extra Postulate will inevitably produce Paradoxes.

It seems unlikely that Nature actually *has* Paradoxes; it seems more likely that only misguided human beings can have Paradoxes. So let us always be parsimonious with Postulates.

SRT uses two numbered Postulates. The First Postulate comes from Galilean Relativity; it says that the Laws of Nature (the differential equations of Physics) are the same in all inertial frames of reference coordinates. The Second Postulate specifies that the speed of light is always the constant c with respect to an observer in any inertial reference frame.

This pair of Postulates *could* constitute *too many* Postulates. Note that the resulting SRT does indeed possess an enormous literature about Paradoxes! In response to the SRT Paradoxes, the present approach dispenses with SRT's Second Postulate, and relies instead on just the First Postulate and the venerable DSB Protocol. Let us see next what it reveals.

ON THE NATURE OF ELECTROMAGNETIC SIGNALS

The governing partial differential equations for an electromagnetic signal are Maxwell's four first-order coupled field equations. Jackson [7] gives Maxwell's equations in modern notation and Gaussian units as:

$$\nabla \bullet \mathbf{B} = 0 \ , \ \nabla \bullet \mathbf{D} = 4\pi\rho \ , \ \nabla \times \mathbf{E} + \frac{1}{c}\partial \mathbf{B} / \partial t = 0 \ , \ \nabla \times \mathbf{H} - \frac{1}{c}\partial \mathbf{D} / \partial t = \frac{4\pi}{c}\mathbf{J}$$

Here $\mathbf{B}$ is the magnetic field and $\mathbf{E}$ is the electric field. The constant $1/c = \sqrt{\varepsilon_0\mu_0}$, where ε_0 is the electric permittivity and μ_0 is the magnetic permeability. In free space, $\mathbf{D} = \varepsilon_0\mathbf{E}$, $\mathbf{H} = \mathbf{B}/\mu_0$, and the charge density ρ and the current density $\mathbf{J}$ are zero. Free space will be the case of interest henceforth in this paper.

It is possible to insert two of the four coupled field equations into the other two, and thereby create two un-coupled, second-order, wave equations:

$$\nabla^2\mathbf{E} - \frac{1}{c^2}\partial^2\mathbf{E} / \partial t^2 = 0 \ , \ \nabla^2\mathbf{B} - \frac{1}{c^2}\partial^2\mathbf{B} / \partial t^2 = 0 \ , \text{ where } 1/c^2 = \varepsilon_0\mu_0$$

Waves were a well-known physical phenomenon even before Maxwell came along. So at the turn of the 20th century it was tempting to think about electromagnetic signals as similar to familiar solutions to these wave equations.

The most familiar type of solution that the un-coupled wave equations support is the infinite plane wave. That name implies a wave that extends to infinity in all three spatial directions, and exists forever in time. In that context, the parameter c is interpreted as the speed of light. Given the history, it was only natural to think of electromagnetic signals as just like infinite plane waves. That is what Einstein did in 1905. [3] It is also what Feynman did half a century later. [4]

Viewed from our vantage point here in the early 21st century, IT actually provides a clear disqualifier for the idea that c is the universal speed for all electromagnetic signals. The problem is that c comes out of the uncoupled wave equations and their infinite-plane-wave solutions. But an infinite plane wave cannot function as a signal, because it cannot convey any information whatsoever! It is no better than an everlasting constant. It is to communication what the bagpipe is to music: an everlasting drone – until some kind of structure is imposed.

Communication of information requires amplitude modulation, or frequency modulation, or on-off switching: some other kind of structure. An infinite plane wave does not have any such structure. The prototypical type of solution that provides the needed structure is the finite-energy electromagnetic pulse. The pulse is what we need to make a proper model for a signal.

The key thing about finite-energy pulses, as apposed to infinite plane waves, is this: under the action of the four coupled field equations, the pulses do not remain pulses; they evolve into spread-out wavelets. This process is irreversible. It does not happen with the two un-coupled wave equations because those equations are second order in time, and so are invariant under time reversal. So no irreversible evolution can be revealed with them.

Let us detail the evolution. Let the word 'pulse' be the short description for a field profile that is rounded on top and sloping down on the sides, and fading gradually to zero; for example, a Gaussian function. The mechanism for the waveform development is that the spatial derivatives applied in Maxwell's equations change the original Gaussian pulse into successively longer wavelets consisting of successively higher order Hermite polynomials multiplying the original Gaussian. As a result, the energy in the wavelet gets more and more spread out along the propagation path.

For a specific illustration, let the field directions be x and y, which makes the propagation direction z. For the record, a Gaussian pulse looks like $G(z) = \dfrac{1}{\sqrt{2\pi}} \exp(-z^2/2)$, and a Hermite polynomial looks like $H_n(z) = \dfrac{1}{G(z)} d^n G(z)/dz^n$, with $n = 1,2,3,4,\ldots$ being the order of the Hermite polynomial.

Figure 1 illustrates a Gaussian pulse and the derived Hermite polynomial that it generates after one complete cycle (*i.e.*, four steps) through Maxwell's first-order coupled field equations. The horizontal axis is the propagation direction, z, and the vertical axis is the field amplitude, $G(z)$ to start with, and $H_4(z) \times G(z)$ after four steps. Series 1 is the input pulse, and Series 2 is the developed wavelet.

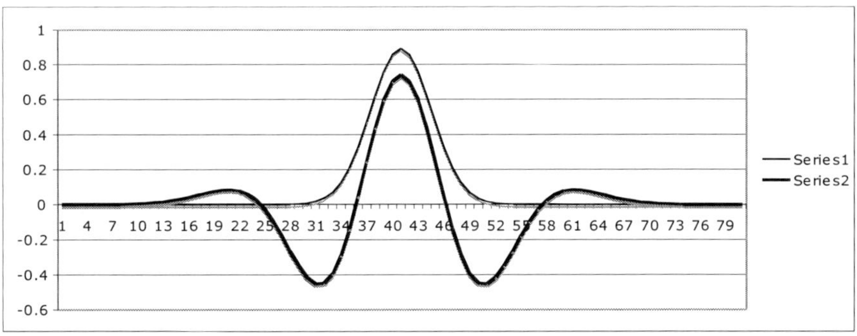

Figure 1. Illustration of signal waveform development.

Now, to solve the stated propagation problem, one can pose a primary pair of pulses in **E** and **B** to guarantee travel, and, for more realism, add a second such pair, offset a quarter cycle in time and perpendicular in space, to model circular polarization, like a real physical photon exhibits. This amounts to four Gaussian functions. But that is not the end of it.

WHAT BOUNDARY CONDITIONS DEMAND

The third part of the Signal model is a pair of propagation boundary conditions: 1) no backflow of energy behind the source, and 2) no overflow of energy beyond the receiver. To guarantee these boundary conditions, we can demand $\mathbf{E} = \mathbf{0}$ at these boundaries. In Fig. 1, the boundaries correspond to the left and right borders. The field amplitude at the boundaries is so small that one cannot tell whether the boundary conditions have been imposed or not. Were the pulse to be near the beginning, or the end, of the path, then the effect of the boundary condition at the beginning, or at the end, of the path would become more visible. Indeed, the propagation process near a boundary is really controlled by the boundary condition at that boundary. The fact that the receiver has this important role in determining the overall evolution of the signal waveform reminds one of a curious aspect of QM: the so-called 'observer effect'. Like QM, the present signal model has an observer effect; namely, the boundary condition at the receiver. But the signal model *also* has an equally important *source* effect; namely, the boundary condition at the source.

We can fulfill both of the required by boundary conditions by matching the signal leaving the source toward the receiver with 1) another fictitious signal going from the source in the opposite direction, with opposite $\mathbf{E}$, to make the total $\mathbf{E}$ field zero at the source, and 2) another fictitious signal approaching the receiver from the opposite direction, with opposite $\mathbf{E}$, so as to make the total $\mathbf{E}$ field zero at the receiver. One can even continue this boundary-fixing process, to clean up each tiny new departure from zero $\mathbf{E}$ at a boundary that each additional fictitious signal creates at the end of the path opposite to the end that it is meant to correct.

Carried to an infinite sum of corrections upon corrections, this is certainly a very complicated picture about an electromagnetic signal. How should we best characterize it in brief? We have all been conditioned to just give the number c for the speed, and presume that number to be relative to the 'observer'; i.e., the receiver of the signal. But clearly, that characterization is inadequate. When we speak of *the* speed of light, we speak way too casually. The little article word '*the*' implies way too much: that there is just *one* light-speed, the number c. But when a light pulse not only travels, but also evolves, there is not a single light-speed *number*. There is a light-speed *function*, depending on both position and time.

It is possible to think about this light-speed function in terms of local Poynting vector $\mathbf{P} = \mathbf{E} \times \mathbf{B}$ and local energy density $S = (E^2 + B^2)/2$. One can speak of a local light speed $c(z) = P(z)/S(z)$. One can speak of a path averaged light speed $<c>= \int P(z)\,dz \Big/ \int S(z)\,dz$. One can include the time variation in $c(z,t) = P(z,t)/S(z,t)$ and $<c>(t) = \int P(z,t)\,dx \Big/ \int S(z,t)\,dz$. One can recognize that $\int S(z,t)\,dz$ is total energy and should be independent of time. One can intuit that while most of the energy is near the source, $<c>(t)$ is c relative to the source, and while most of the energy is near the receiver, $<c>(t)$ is c relative to the receiver. But what about the great bulk of the scenario, during which the energy is *not* near either the source *or* the receiver?

We need a new and more appropriate statement to use for the DSB boundary conditions. The above statements can be characterized as 'one number, two references'. Instead, let us consider the reverse: 'one reference, two numbers'. First, look at the propagation scenario overall, and take note of an important symmetry property that it exhibits. There exists a temporal midpoint of the scenario, and at that moment, the waveform is symmetric about the spatial midpoint of the propagation path. Figure 1 corresponds to this mid-point situation, with the source on the left side of the Figure, and the receiver on the right side.

Consider that it cannot matter very much how the receiver *previously* moved before that temporal midpoint, and it cannot matter very much how the source will *subsequently* move after that temporal midpoint. The signal is roughly like a baton, handed off from source to receiver at that temporal midpoint. The length of that baton defines the distance traveled from the source to the receiver, and the direction of the baton defines the angular position of the source in relation to the receiver.

A good Rule of Thumb is the following: Whenever we cannot avoid speaking of light speed as a number c relative to something, let that something be the spatial midpoint of that imaginary baton. This statement is different from Einstein's Second Postulate [3] that c be always relative to the receiver, and different from Ritz's counter proposal [7] that c be always relative to the source, *and* different from the first option described above: 'one number, two references'. When we speak of c relative to the spatial midpoint of the propagation scenario, that characterization amounts to *not* c relative to the source and *not* c relative to the receiver. It is the realization of the second option: 'one reference, two numbers'.

To be sure, the Rule of Thumb offered here is an approximation. But absent large accelerations, it is a very good approximation. So let us see next what new understanding it produces.

REALITY *VS.* INTERPRETATION

Consider a simple scenario with steady relative motion between an observer and a target. Consider a real-world problem: estimating the speed of a moving target relative to an observer. One way or another, there must exist *two* signals, the first one outgoing from the observer to the target, and the second one returning from the target to the observer. So for one signal, the observer is the source and the target is the receiver, and for the other signal, the target is the source and the observer is the receiver.

Each of the two signals has two phases: expansion from its source, and collapse to its receiver. Thus this scenario has altogether *four* parts. Call them Parts 1 through 4.

If the motion is retreat at speed V, then having speed c relative to the scenario midpoint means we have $c+V/2$ relative to the source, and $c-V/2$ relative to the receiver (presuming for now that $V<2c$). If the motion is approach at speed V, we have the opposite case, $c-V/2$ relative to the source (again presuming $V<2c$), and $c+V/2$ relative to the receiver.

To keep track, it helps to retire the vague concept of '*the* speed of light', and speak more specifically of 'the speed of the leading front of the developing wavelet' and 'the speed of the trailing end of the developing wavelet'. During Part 1, the outgoing expansion from the observer, the leading front moves at $2c$ relative to the source, or $2c-V$ relative to the target. During Part 2, the collapse to the target, the trailing end moves at $2c$ relative to the target, or $2c+V$ relative to the source. Then during Part 3, the return expansion, the leading front moves at $2c$ relative to the target, or $2c-V$ relative to the observer. Finally, during Part 4, the collapse to the observer, the trailing end moves at $2c$ relative to the observer, $2c+V$ relative to the target.

Suppose the initial separation between source and target is L_0, and the relative speed between source and target is V. For the record, if $V \geq 2c$ the linear-motion scenario stops right here: either the signal from the observer could never reach its target, or the return signal from the target could never arrive back to the observer. So we continue the present analysis under the assumption that $V < 2c$.

1) After Part 1 (expansion from the observer) the separation is:

$$L_1 = L_0 + L_0 \times (V / 2c) + L_0 \times (V / 2c)^2 + ... = L_0 / (1 - V / 2c) .$$

2) After Part 2 (collapse to the target) the separation is:

$$L_2 = L_1 \times (1 + V / 2c) = L_0 \times (1 + V / 2c) / (1 - V / 2c) .$$

3) After Part 3 (re-expansion from the target) the separation is:

$$L_3 = L_2 + L_2 \times (V / 2c) + L_2 \times (V / 2c)^2 + ... = L_2 / (1 - V / 2c)$$
$$= L_0 \times (1 + V / 2c) / (1 - V / 2c)^2 .$$

4) After Part 4 (re-collapse to the observer) the separation is:

$$L_4 = L_3 \times (1 + V / 2c) = L_0 \times (1 + V / 2c)^2 / (1 - V / 2c)^2 .$$

Compare this case to the case where the scenario follows the Einstein Postulate. There would be only two parts: outbound travel at c relative to the target, followed by return travel at c relative to the observer. And the scenario could not play out if $V \geq c$. So we continue the analysis under the assumption that $V < c$.

1) After Part 1 (travel from the observer to the target) the separation would be:

$$L_1 = L_0 / (1 - V / c) .$$

2) After Part 2 (return travel from the target to the observer) the separation would be:

$$L_2 = L_1 \times (1 + V / c) = L_0 \times (1 + V / c) / (1 - V / c) .$$

Observe that the length expansion with the Einstein Postulate is less than that with the four steps.

For the sake of completeness, consider also the case where the scenario follows the Ritz Proposal. Again there would be only two parts, but they would be slightly different, because the outbound trip would be at c relative to the observer, and the return trip would be at c relative to the target.

1) After Part 1 (travel from the observer to the target) the separation would be:

$$L_1 = L_0 \times (1 + V / c) \ .$$

2) After Part 2 (return travel from the target to the observer) the separation would be:

$$L_2 = L_1 \big/ (1 - V / c) = L_0 \times (1 + V / c) \big/ (1 - V / c) \ .$$

So the length expansion with the Ritz Postulate is similarly less than that with the four steps that make the present model.

These discrepancies create a problem. When the actual function is $(1 + V / 2c)^2 \big/ (1 - V / 2c)^2$, while the believed model is $(1 + V / c) \big/ (1 - V / c)$, the only way to make the believed model match the actual function is to believe in a reduced value of speed, some $v < V$. To produce the required match between the actual function and the believed model, one needs $v = V \big/ (1 + V^2 / 4c^2)$:

$$\frac{1 + v / c}{1 - v / c} = \frac{1 + (V / c) \big/ (1 + V^2 / 4c^2)}{1 - (V / c) \big/ (1 + V^2 / 4c^2)} = \frac{1 + V / c + V^2 / 4c^2}{1 - V / c + V^2 / 4c^2} \equiv \frac{(1 + V / 2c)^2}{(1 - V / 2c)^2} \ .$$

Given the distortion $v = V \big/ (1 + V^2 / 4c^2)$, other distortions quickly follow. If the data being collected concerns the length of a measuring rod, then the rod appears to be shortened in the travel direction, or if the data concerns the image of a clock face, then the clock appears to run slow, and so on.

Observe that the function $\frac{1+v/c}{1-v/c}$ blows up for $v=c$, which is surprising. Further observe that the function $\frac{1+v/c}{1-v/c}$ changes sign for $v>c$, which makes no physical sense at all. Because of these facts, it has long been thought that Nature has a physical speed limit $v<c$. But the idea that Nature would have a speed limit is also generally recognized as a Big Mystery.

The present analysis suggests that little v is just a convention, an artifact accepted from a somewhat faulty mathematical model, and *not* a fact established from unbiased observation or uncontested theory. Situations like this are well known in Engineering Science: in controlling the behavior of a physical system, a mathematical model that strays too far from the physical reality will lead to strange, and sometimes catastrophic, consequences.

But what about big V? Observe that the function $\frac{(1+V/2c)^2}{(1-V/2c)^2}$ blows up for $V=2c$. Does this fact mean that V has a limit $V<2c$ corresponding to $v<c$? No, it does not! What V has is a limit on *observation*: speeds $V>2c$ may indeed exist, but for linear motion, they simply cannot be observed using light signals.

In circular motion, however, the situation is different. Speed $V>2c$ may be observable, and it may, in fact, have been observed. Kuligin et al., [9] tell a story suggesting this conclusion. Let a linear particle accelerator deliver a bolus of charged particles into a magnetic field, where charges follow a circular orbit.

The resulting situation sometimes *looks* as if the charge bolus breaks into fragments. But the observation *also* allows the alternative interpretation that the charge bolus remains intact, but travels *much faster* than SRT allows. So in a time interval for which we expect to see the charge bolus go by only once, we actually see it go by twice, or even more times!

Now for the sake of completeness, observe that the relationship $v=V/(1+V^2/4c^2)$ implies a quadratic equation for V as a function of v:

$$(v/4c^2)V^2 - V + v = 0 \ .$$

This equation has two solutions, $V = \frac{1 \pm \sqrt{1 - v^2 / c^2}}{v / 2c^2}$. For small v, $V = \frac{1 \pm (1 - v^2 / 2c^2)}{v / 2c^2}$, which evaluates to $V = \mp v$ and $V = \frac{2 - v^2 / 2c^2}{v / 2c} \rightarrow \frac{4c^2}{v}$. That is to say, V evaluates to one familiarly small value, and one unfamiliarly huge value!

POTENTIALS AND FIELDS

Let us now proceed to review SRT in a different and more important matter; namely, electromagnetic potentials and fields. These are functions of position $\mathbf{r}$ and time t where some receiver can sense them. The potentials are the scalar potential $\Phi(\mathbf{r}, t)$ and the vector potential $\mathbf{A}(\mathbf{r}, t)$. Through spatial and temporal derivatives, they give rise to the electric field $\mathbf{E}(\mathbf{r}, t)$ and the magnetic field $\mathbf{B}(\mathbf{r}, t)$:

$$\mathbf{E}(\mathbf{r}, t) = -\nabla\Phi(\mathbf{r}, t) - \frac{1}{c} \partial\mathbf{A} / \partial t \ , \ \mathbf{B}(\mathbf{r}, t) = \nabla \times \mathbf{A}(\mathbf{r}, t).$$

The problem of characterizing the electromagnetic potentials and fields registered at a receiver predates Einstein, and the Assumption that all researchers made about it was so ingrained that it was not even recognized as being an Assumption, until Einstein called it out as his Second Postulate in [3]. So he is to be much credited for doing that.

The sources generally cited for the early (1898 to 1901) work on potentials and fields are A. Liénard [10] and E. Wiechert [11]. Although they worked at about the same time, they worked separately. And they got the same results – as did everyone else working on the problem around that time.

The Liénard-Wiechert results are given in every standard EM book, and are summarized again here, for review and subsequent further discussion. The standard scalar and vector potentials are:

$$\Phi(\mathbf{r}, t) = e\left[1 / \kappa R\right]_{\text{retarded}} \text{ and } \mathbf{A}(\mathbf{r}, t) = e\left[\boldsymbol{\beta} / \kappa R\right]_{\text{retarded}} ,$$

where $\kappa = 1 - \mathbf{n} \bullet \boldsymbol{\beta}$, $\boldsymbol{\beta}$ is source velocity normalized by c, and $\mathbf{n} = \mathbf{R} / R$ (a unit vector), and $\mathbf{R} = \mathbf{r}_{\text{source}}(t - R / c) - \mathbf{r}_{\text{receiver}}(t)$ (an implicit definition for the terminology 'retarded').

The LW fields obtained from those potentials are then:

$$\mathbf{E}(\mathbf{r},t) = e\left\{ \frac{1}{\kappa^3 R^2} (\mathbf{n} - \boldsymbol{\beta})(1 - \beta^2) + \frac{1}{c\kappa^3 R} \mathbf{n} \times \left[(\mathbf{n} - \boldsymbol{\beta}) \times (d\boldsymbol{\beta} / dt) \right] \right\}_{\text{retarded}}$$

and $\mathbf{B}(\mathbf{r},t) = \mathbf{n}_{\text{retarded}} \times \mathbf{E}(\mathbf{r},t)$.

The $1 / R^2$ fields are called Coulomb-Ampere fields. Observe the direction of the Coulomb electric field: $(\mathbf{n} - \boldsymbol{\beta})_{\text{retarded}}$. If $\boldsymbol{\beta}$ does not change much over the total field propagation time, then $(\mathbf{n} - \boldsymbol{\beta})_{\text{retarded}}$ is virtually indistinguishable from $\mathbf{n}_{\text{present}}$.

The $1 / R$ fields are radiation fields, and together they make a Poynting vector (energy flow per unit area per unit time):

$$\mathbf{P} = \frac{c}{4\pi} \mathbf{E}_{\text{radiative}} \times \mathbf{B}_{\text{radiative}} = \frac{c}{4\pi} \mathbf{E}_{\text{radiative}} \times \left(\mathbf{n}_{\text{retarded}} \times \mathbf{E}_{\text{radiative}} \right) = \frac{c}{4\pi} \left(E_{\text{radiative}} \right)^2 \mathbf{n}_{\text{retarded}}$$

The Poynting vector lies along $\mathbf{n}_{\text{retarded}}$.

So the Coulomb field and the Poynting vector are arriving to the observer from distinctly different directions: approximately $\mathbf{n}_{\text{present}}$ *vs.* $\mathbf{n}_{\text{retarded}}$. This result does *not* look right: it looks like advance information being provided on one channel, but not the other, of the two information channels: the Coulomb signal and radiation signal.

Nevertheless, everyone working at that time adopted the same Assumption, and endorsed the same result. The Liénard-Wiechert Assumption, and later the Einstein Postulate, was: the speed of light is always c with respect to the receiver of the light.

Now consider the same problem using the new and more nuanced definition for the light speed reference: the 'baton' handed over from the source to the receiver at the temporal midpoint of the propagation scenario. Observe that the line that connects the source position to the receiver position at the temporal midpoint of the scenario defines both the distance and the direction that the energy must travel in order to achieve the particular transmission of a signal from the source to the receiver that is being modeled. This specification implies that we need potentials and fields to be, not *retarded*, but ***half***-*retarded*. Now the potentials become:

$$\Phi(\mathbf{r},t)=e\left[1/\kappa R\right]_{\mathbf{half}-\text{retarded}} \text{ and } \mathbf{A}(\mathbf{r},t)=e\left[\boldsymbol{\beta}/\kappa R\right]_{\mathbf{half}-\text{retarded}} .$$

The fields become:

$$\mathbf{E}(\mathbf{r},t)=e\left\{\frac{(\mathbf{n}-\boldsymbol{\beta})(1-\beta^2)}{\kappa^3R^2}+\frac{\mathbf{n}}{c\kappa^3R}\times\left[(\mathbf{n}-\boldsymbol{\beta})\times d\boldsymbol{\beta}/dt\right]\right\}_{\mathbf{half}-\text{retarded}}$$

and $\mathbf{B}(\mathbf{r},t)=\mathbf{n}_{\mathbf{half}-\text{retarded}}\times\mathbf{E}(\mathbf{r},t)$.

Observe the direction of the Coulomb $\mathbf{E}$ field:

$$(\mathbf{n}-\boldsymbol{\beta})_{\mathbf{half}\text{-retarted}}\approx(\mathbf{n}_{\text{present}})_{\mathbf{half}\text{-retarted}}\triangleq\mathbf{n}_{\mathbf{half}\text{-retarted}} .$$

The Poynting vector becomes:

$$\mathbf{P}=\frac{c}{4\pi}\mathbf{E}_{\text{radiative}}\times\mathbf{B}_{\text{radiative}}$$

$$=\frac{c}{4\pi}\mathbf{E}_{\text{radiative}}\times\left(\mathbf{n}_{\mathbf{half}-\text{retarded}}\times\mathbf{E}_{\text{radiative}}\right)=\frac{c}{4\pi}(E_{\text{radiative}})^2\mathbf{n}_{\mathbf{half}-\text{retarded}} .$$

This vector too lies along $\mathbf{n}_{\mathbf{half}-\text{retarded}}$. That is, the Coulomb field and the Poynting vector are now reconciled to the *same* direction, instead of conflicting with each other. Does this reconciliation really matter? Indeed it does. With this one little adjustment, we are far better prepared to understand the next problem at hand: the behavior of an atom.

MODELING AN ATOM

QM as we presently know it was founded, in part, to avoid a seeming problem with the Hydrogen atom viewed classically: the electron in its circular orbit must accelerate, and so must radiate away its orbit energy, and so must collapse into the nucleus. That circumstance forces us to inject yet another Postulate; namely, that classical physics is not sufficient at the level of atoms.

But with the direction of the Coulomb field being $\mathbf{n}_{\textbf{half}-\text{retarded}}$, there is a tiny tangential component of Coulomb force aligned with the orbit velocity. So there is a torque on the atom, and the torque pumps energy into the atom, and that process can work to balance the energy loss due to radiation. Here are some relevant details:

Let the masses of the electron and the proton be m_{e} and m_{p}. Note that $m_{\text{e}} << m_{\text{p}}$, but m_{p} is *not* infinite.

Let the orbit radii of the electron and the proton be r_{e} and r_{p}. Note that $r_{\text{p}} << r_{\text{e}}$, but r_{p} is *not* zero.

Let the charges on the electron and the proton be $-e$ and $+e$.

The magnitude of the nominally attractive force within the atom is $F = e^2 \big/ (r_{\text{e}} + r_{\text{p}})^2$.

Let the orbit frequency be Ω. The orbit speed of the electron is $v_{\text{e}} = r_{\text{e}}\Omega$ and that of the proton is $v_{\text{p}} = r_{\text{p}}\Omega$.

The magnitude of the tiny tangential force on the electron is $F_{\text{e}} = Fv_{\text{p}} \big/ 2c = Fr_{\text{p}}\Omega \big/ 2c$.

The magnitude of the tiny tangential force on the proton is $F_{\text{p}} = Fv_{\text{e}} \big/ 2c = Fr_{\text{e}}\Omega \big/ 2c$.

The magnitudes of the torques on the electron and on the proton are $T_{\text{e}} = r_{\text{e}}F_{\text{e}}$ and $T_{\text{p}} = r_{\text{p}}F_{\text{p}}$, both equal to $Fr_{\text{e}}r_{\text{p}}\Omega \big/ 2c$.

The total torque on the electron-proton system is $T_{\text{total}} = T_{\text{e}} + T_{\text{p}} = 2T = Fr_{\text{e}}r_{\text{p}}\Omega^2 \big/ c$.

The torque power delivered to the system is $P_{\text{torque}} = T\Omega = Fr_e r_p \Omega^2 \big/ c$.

The torque power delivered to the system is $P_{\text{torque}} = T\Omega = Fr_e r_p \Omega^2 \big/ c$.

The squared orbit frequency is determined from either $F = m_{\text{e}}r_{\text{e}}\Omega^2$ or $F = m_{\text{p}}r_{\text{p}}\Omega^2$.

The more convenient of the two options is $\Omega^2 = F / m_e r_e$. With that expression, the approximation $r_e \approx r_e + r_p$ yields:

$$P_{\text{torque}} = F r_e r_p \Omega^2 / c = F^2 r_p / m_e c = \left(e^4 r_p / m_e c\right) / \left(r_e + r_p\right)^4 \approx \left(e^4 / m_p c\right) / \left(r_e + r_p\right)^3$$

The P_{torque} is actually quite large, and so it changes the whole emphasis of worry concerning the Hydrogen atom. The question becomes, not why does the Hydrogen atom not radiate and collapse to death, but rather why does the Hydrogen atom not torque itself up and expand way beyond its known size?

The fact is: there exists much, much more radiation than was previously worried about, and it is enough to produce the proper balance between radiation and torquing. The extra radiation arises from finite signal propagation speed, which results in circular motion of the center of mass of the Hydrogen atom. This in turn produces Thomas rotation [12], and thereby scales up by a factor of 2 the overall rotation rate generating the radiation, which increases the radiation power by a factor of 2^4 .

But what should one say about that center-of-mass circular motion? In Newtonian physics, where signal propagation speed is infinite, there is no such thing. In Maxwell physics, the emphasis is on fields, and the responses of individual charges, but not as much on the responses of whole charge systems, such as atoms. So the issue doesn't come up there. In Einstein's relativity physics, the emphasis is often on the observers of events more than on the events themselves. System center-of-mass circulation seems not to come up, although Thomas rotation does.

Thomas rotation is generally believed to be a result of the properties of Lorentz transformations, and hence of SRT. That is the belief because one can think of Lorentz transformations, not only in the usual, passive sense, as conversion from an observer in one inertial coordinate frame to another observer in another inertial coordinate frame, but also in the active sense, as the application of a 'boost' in velocity; that is, as an acceleration, the result of a physical force. A series of non-co-linear boosts does indeed produce Thomas rotation.

But Thomas rotation does arise, not just from Lorentz transformations, but also from Galilean Transformations. That fact can be demonstrated in detail as follows:

For simplicity, let all motion be in the x, y plane. The scenario begins at time coordinate ct_0 with one of the particles, say the electron, at rest at spatial coordinates x_0, y_0. Let an attraction from another particle act in the x direction. Let an increment of velocity $\Delta\mathbf{V}_x = \Delta V$ be imposed, and let an increment of time Δt elapse. The coordinates of the electron then become:

$$t_1 = c(t_0 + \Delta t)\ ,\ x_1 = x_0 + \Delta V \Delta t\ \text{, and}\ y_1 = y_0\ .$$

Now let an attraction from another particle act in the y direction. Let an increment of velocity $\Delta\mathbf{V}_y = \Delta V$ be imposed, and let another increment of time Δt elapse. The coordinates of the electron then become:

$$t_2 = c(t_0 + \Delta t + \Delta t),\ x_2 = x_0 + \Delta V(\Delta t + \Delta t) = x_0 + 2\Delta V \Delta t\ \text{, and,}$$
$$y_2 = y_0 + \Delta V \Delta t\ .$$

Observe that, if the Galilean velocity boosts had been applied in the opposite order, then the ending coordinates of the electron would have been:

$$t_2 = c(t_0 + \Delta t + \Delta t)\ ,\ x_2 = x_0 + \Delta V \Delta t\ \text{, and}\ y_2 = y_0 + 2\Delta V \Delta t\ .$$

Observe that the squared incremental length changes have the same magnitude either way:

$$(2\Delta V \Delta t)^2 + (\Delta V \Delta t)^2 \equiv (\Delta V \Delta t)^2 + (2\Delta V \Delta t)^2 = 5(\Delta V \Delta t)^2$$

That fact means the two possible sequences of Galilean boost applications differ only by a rotation. That means each one individually contains a rotation equal to half that total angle difference. This is the Thomas rotation.

Without the Thomas rotation, the radiation power from the Hydrogen atom would be:

$$P_{\text{radiated}} = \frac{2e^2}{3c^3} a_{\text{e}}^2 = \left(2e^6 / m_{\text{e}}^2\right) / 3c^3 (r_{\text{e}} + r_{\text{p}})^4\ .$$

With the Thomas rotation included, the radiation power from the Hydrogen atom is:

$$P_{\text{total radiated}} = \underline{2}^4 \frac{2e^2}{3c^3} a_e^2 = \left(\underline{2}^5 e^6 / m_e^2\right) \Big/ 3c^3 (r_e + r_p)^4 \ .$$

The value of the separation $r_e + r_p$ for which $P_{\text{total radiated}} = P_{\text{torque}}$ is:

$$r_e + r_p = 32 m_p e^2 \big/ 3 m_e^2 c^2 = 5.5 \times 10^{-9} \text{ cm} \ .$$

Recall that in the traditional approach to QM, $r_e + r_p = h^2 \big/ 4\pi^2 \mu e^2$, where μ is the reduced mass, defined by $\mu^{-1} = m_e^{-1} + m_p^{-1}$, and very nearly equal to m_e , and h is Planck's constant, 6.626176×10^{-34} Joule-sec, a fundamental constant given by Nature.

The present approach does not use Planck's constant as an input. Instead, it gives an estimate of Planck's constant as an output:

$$h = \sqrt{4\pi^2 \mu e^2 (r_e + r_p)} \approx \sqrt{4\pi^2 m_e e^2 32 m_p e^2 \big/ 3 m_e^2 c^2}$$
$$\approx \frac{\pi e^2}{c} \sqrt{128 m_p / 3 m_e} \approx 6.77 \times 10^{-34} \text{Joule-sec.}$$

This estimate can be improved by taking due account of the fact that the sines of small angles are not exactly equal to the angles themselves, and the cosines of small angles are not exactly equal to unity. The sine corrections are third order in angle, while the cosine corrections are second order in angle, which is more significant. Including those corrections reduces the radiation power slightly, and so reduces the solution $r_e + r_p$ slightly, and so reduces the estimate of h slightly – a step in the right direction.

What is the broader significance of this analysis of the Hydrogen atom? It illustrates an approach that involves no founding Postulate, and requires no elaborate computation; it is just Algebra. The approach can help, not just with the Hydrogen atom, but also with *all* atoms. And it can help, not just with single atoms, but also with complex *systems* consisting of multiple atoms.

I see two major areas of study involving complex systems consisting of multiple atoms. One area is Spectroscopy. It has been suggested that *all* light generation may be similar to laser-light generation. [13] Laser light arises, *not* from *single* atoms, but rather from *multiple* atoms. The multiple atoms constitute a complex system, and the complex system emits *multiple* photons upon transition into other less complex systems. That idea can account for the complexity of observable spectra, more plausibly I believe, than can the traditional idea of single atoms possessing multiple possible excited states. This area of investigation is certainly interesting, but it is not obviously relevant to CF.

The other area of study involving complex systems consisting of multiple atoms is Chemistry. I believe a convenient approach to Chemistry is truly vital for understanding and engineering CF. The next Section discusses a new approach to Chemistry that offers a computational advantage over existing approaches.

THE BASICS OF ALGEBRAIC CHEMISTRY

The title of my book [1] Algebraic Chemistry (AC) reflects the fact that the technique has no integrals or other complicated math operations that would demand capabilities beyond those of a hand calculator. The worst operation is square root. So the AC approach looked promising for quick application to problems such as CF.

The fundamental idea behind AC is that all atoms share some similarities with Hydrogen atoms: 1) They have a nucleus that is similar to a proton, but scaled up to nuclear charge Z and nuclear mass M ; 2) They have a population of electrons that is not entirely unlike a single electron; i.e., an interacting community that is somewhat coherent, and somewhat like one *big* electron orbiting the nucleus; 3) It is possible for the electron count to be different from the nuclear charge. This last possibility is what characterizes ions, and thereby creates all of Chemistry.

The initial observation from which the fundamental idea developed is this: neutrons are a complicating factor for Chemistry. They create a huge variety of slightly different atoms, with slightly different properties. But the model for the Hydrogen atom from the last Section suggests a way to reduce the complexity and see an underlying pattern.

Recall the formula for Hydrogen atom charge separation:

$$r_e + r_p = 32 m_p e^2 / 3 m_e^2 c^2 = 5.5 \times 10^{-9}$$

The separation $r_e + r_p$ scales with the proton mass, m_p. That means the basic orbit energy of the Hydrogen atom scales with $1/m_p$. It suggests that the basic orbit energy of an arbitrary atom may scale with Z/M, where Z is the nuclear charge, and M is the nuclear mass. Therefore, scaling the measured first-order ionization potentials for all elements by the inverse factor, M/Z, may reveal some previously hidden pattern.

We find that the scaled first-order ionization potentials (called $IP_{1,Z}$) do indeed fall into a *very* regular pattern: the rise on *every* period in the Periodic Table is *exactly the same factor*, $7/2$. The consistency of the $7/2$ rise on every period suggests that the *IP* values are not just about individual elements; they are following a pattern that relates *all* elements, and because of the pattern, each $IP_{1,Z}$ contains information about all *other* elements. That is, $IP_{1,Z}$ embodies *population generic information.*

The breakdown that suggests itself is this: each $IP_{1,Z}$ contains a universal baseline contribution $IP_{1,1}$ about interaction between the nucleus and the population of electrons as a whole, and, for all elements beyond Hydrogen, a contribution $\Delta IP_{1,Z}$ about interactions just among the electrons by themselves.

The electron-electron contribution can be dramatic. Consider Helium: $\Delta IP_{1,2}$ is *huge*, meaning that two electrons make a very strong bond between themselves. And consider Lithium: $\Delta IP_{1,3}$ is negative, meaning that two electrons actively work together to try to exclude a third electron.

Over the periods, there is obvious detail about the electron-electron interactions. Within each period, there are obvious sub-periods keyed to the nominal angular-momentum quantum number that is being filled. Plotted on a log scale, all sub-period rises are straight lines. The slopes all appear to be rational fractions. Table 1 displays these fractions.

Table 1. Ionization Potential Rises in the Periodic Table

period	N	l	fraction	l	fraction	l	fraction	l	fraction
1	1	0	1						
2	2	0	1/2	1	3/4				
3	2	0	1/2	1	3/4				
4	3	0	1/4	2	5/18	1	2/3		
5	3	0	1/4	2	5/18	1	2/3		
6	4	0	1/4	3	7/48	2	5/16	1	9/16
7	4	0	1/4	3	7/48	2	5/16	1	9/16

Observe that a non-traditional parameter N is included in the display. For $l > 0$, this N makes it possible to write a simple formula for the fraction needed:

$$\text{fraction} = \left[(2l+1)/N^2\right]\left[(N-l)/l\right] .$$

Also, N makes it possible to characterize the lengths of all periods in the Periodic Table: the period length is always $2N^2$.

All this numerical regularity strongly suggests that there really is a reliable pattern here, and that we can reasonably seek to exploit it for calculating what events to expect in Chemical interactions. Here is the first exploitation that suggests itself: Given first-order ionization potentials of many elements, we can make reasonable estimates of the additional energy required to remove a second electron from each, and then a third, and so on.

Please note: estimating the energy to remove a second electron, or a third, and so on, is *not* the same thing as estimating the so-called 'second-order ionization potential', 'third-order ionization potential', and so on. Look up those energies, and you will see that they are very large. The implication is that the big numbers are about events that are very *violent*: ripping two, or three, or more, electrons off an atom *all at once*. This isn't what happens in ordinary every-day lab-bench Chemistry: most ionization events occur *gently*, one at a time.

Just as for the first-order ionization potentials, there is a great deal of numerical regularity in the higher-order ionization potentials. You can read about it, and view a big display Figure about it, in Chapter 1 of [1]. But the subject of the present paper does not require all this information. That is because violent events occur only when some really heavy-duty equipment, such as a HF reactor, is involved. That is not the case for CF. The basic chemistry of water, involving only first-order ionization potentials, is enough for CF. [14]

The following development represents a simplification over that in [1]: it consolidates formulae about the one-by-one removal or addition of a number of electrons, into formulae about cumulative removal or addition of that number of electrons. This consolidation does not change end results, but it does save some computation work along the way.

The formulae all involve increments of energy identified as work W and heat H. Work is energy required to move electrons from one atom to another atom. Heat is about internal readjustments that electron populations within the atoms then make in response.

n an atom of the element with nuclear charge Z, the work to remove one electron from the nuclear field is:

$$W_{\text{removing e}_1\text{ from the neutral atom}} = IP_{1,1}(Z / M_Z) \ .$$

This is energy that has to be supplied to get the removal process started: energy to get over a potential wall. Note the factor of Z / M. It restores *population-generic* information, IP 's, to *element-specific* information.

The subsequent re-adjustment among electrons within the atom involves heat:

$$H_{\text{removing e}_1\text{ from the neutral atom}} = (\Delta IP_{1,Z} - \Delta IP_{1,Z-1})(Z / M_Z) \ .$$

Note: for the special case $Z = 1$, $\Delta IP_{1,Z-1} = \Delta IP_{1,0}$ does not exist, and $\Delta IP_{1,Z} = \Delta IP_{1,1} = 0$. And for $Z = 2$, and $\Delta IP_{1,Z-1} = \Delta IP_{1,1} = 0$.

Thus the total energy involved in removing one electron from an atom with nuclear charge Z is:

$$(W + H)_{\text{removing e}_1\text{ from the neutral atom}} = \left[IP_{1,1}Z + \Delta IP_{1,Z}Z - \Delta IP_{1,Z-1}(Z-1) \right] \big/ M_Z$$

Now suppose that we remove a second electron. The work that has to be supplied because of the nuclear field is:

$$W_{\text{removing e}_2 \text{ after e}_1} = IP_{1,1}\sqrt{Z(Z-1)}\Big/M_Z \ .$$

Observe the factor of $\sqrt{Z(Z-1)}$, instead of just Z; it takes account of the difference between nuclear charge and electron count.

The heat then involved in electron readjustment is:

$$H_{\text{removing e}_2 \text{ after e}_1} = \Delta IP_{1,Z-1}(Z-1)/M_Z - \Delta IP_{1,Z-2}(Z-2)/M_Z \ .$$

The total energy involved in this step is then:

$$(W+H)_{\text{removing e}_2 \text{ after e}_1} =$$
$$IP_{1,1}\sqrt{Z(Z-1)}\Big/M_Z + \Delta IP_{1,Z-1}(Z-1)/M_Z - \Delta IP_{1,Z-2}(Z-2)/M_Z$$

Now put both steps together. The grand total energy involved in removing two electrons is then:

$$(W+H)_{\text{removing e}_1 \ \& \ \text{e}_2} = I$$
$$P_{1,1}\left[Z+\sqrt{Z(Z-1)}\right]\Big/M_Z + \left[\Delta IP_{1,Z}Z - \Delta IP_{1,Z-2}(Z-2)\right]\Big/M_Z \ .$$

Note the cancellation of $\Delta IP_{1,Z-1}(Z-1)/M_Z$ terms; such cancellations save some computation in getting to higher orders of ionization. The grand total energy involved in removing three electrons will be just:

$$(W+H)_{\text{removing e}_1, \text{e}_2, \ \& \text{e}_3} =$$
$$P_{1,1}\left[Z+\sqrt{Z(Z-1)}+\sqrt{Z(Z-2)}\right]\Big/M_Z + \left[\Delta IP_{1,Z}Z - \Delta IP_{1,Z-3}(Z-3)\right]\Big/M_Z \ ,$$

and so on, in that pattern.

Now let us go in the other direction, *adding* electrons. The work involved in *adding* one electron into the nuclear field of the atom with nuclear charge Z is:

$$W_{\text{adding e}_1 \text{ to neutral atom}} = -IP_{1,1}\sqrt{Z(Z+1)}\Big/M_Z \ .$$

The heat involved in then re-adjusting the electron population is:

$$H_{\text{adding e}_1 \text{ to neutral atom}} = -\Delta IP_{1,Z+1}(Z+1) / M_Z + \Delta IP_{1,Z} Z / M_Z \ .$$

so that

$$(W+H)_{\text{adding e}_1 \text{ to neutral atom}} = -\left[IP_{1,1}\sqrt{Z(Z+1)} + \Delta IP_{1,Z+1}(Z+1) - \Delta IP_{1,Z} Z \right] / M_Z$$

If we add a second electron to the singly charged ion, it will require additional work:

$$W_{\text{adding e}_2 \text{ after e}_1} = -IP_{1,1}\sqrt{Z(Z+2)} / M_Z \ .$$

And it will cause a further heat adjustment:

$$H_{\text{adding e}_2 \text{ after e}_1} = -\Delta IP_{1,Z+2}(Z+2) / M_Z + \Delta IP_{1,Z+1}(Z+1) / M_Z \ .$$

Putting both steps together, the grand total energy involved in adding two electrons is:

$$(W+H)_{\text{adding e}_1 \text{ \& e}_2}$$
$$= -IP_{1,1}\left[\sqrt{Z(Z+1)} + \sqrt{Z(Z+2)} \right] / M_Z + \left[\Delta IP_{1,Z} Z - \Delta IP_{1,Z+2}(Z+2) \right] / M_Z .$$

Similarly, the total energy involved in adding three electrons is:

$$(W+H)_{\text{adding e}_1, \text{e}_2 \text{ \& e}_3} =$$
$$-IP_{1,1}\left[\sqrt{Z(Z+1)} + \sqrt{Z(Z+2)} + \sqrt{Z(Z+3)} \right] / M_Z + \left[\Delta IP_{1,Z} Z - \Delta IP_{1,Z+3}(Z+3) \right] / M_Z .$$

And so it goes on, in that pattern.

The data needed for study of Water Chemistry and CF are a very small subset of the data for all the elements known, or anticipated, to exist. The full data set is available in [1]. But here we need at most the data for the first ten elements. These data, expressed in electron volts, eV, are as follows:

Hydrogen: $Z = 1$, $M = 1.008$, $IP_{1,1} = 14.250\,\text{eV}$, $\Delta IP_{1,1} = 0\,\text{eV}$

Helium: $Z = 2$, $M = 4.003$, $IP_{1,2} = 49.875\text{eV}$, $\Delta IP_{1,2} = 35.625\text{eV}$

Lithium: $Z = 3$, $M = 6.941$, $IP_{1,3} = 12.469\text{eV}$, $\Delta IP_{1,3} = -1.781\text{eV}$

Beryllium: $Z = 4$, $M = 9.012$, $IP_{1,4} = 23.327\text{eV}$, $\Delta IP_{1,4} = 9.077\text{eV}$

Boron: $Z = 5$, $M = 10.811$, $IP_{1,5} = 17.055\text{eV}$, $\Delta IP_{1,5} = 2.805\text{eV}$

Carbon: $Z = 6$, $M = 12.011$, $IP_{1,6} = 21.570\text{eV}$, $\Delta IP_{1,6} = 7.320\text{eV}$

Nitrogen: $Z = 7$, $M = 14.007$, $IP_{1,7} = 27.281\text{eV}$, $\Delta IP_{1,7} = 13.031\text{eV}$

Oxygen: $Z = 8$, $M = 15.999$, $IP_{1,8} = 27.281\text{eV}$, $\Delta IP_{1,8} = 13.031\text{eV}$

Fluorine: $Z = 9$, $M = 18.998$, $IP_{1,9} = 34.504\text{eV}$, $\Delta IP_{1,9} = 20.254\text{eV}$

Neon: $Z = 10$, $M = 20.180$, $IP_{1,10} = 43.641\text{eV}$, $\Delta IP_{1,10} = 29.391\text{eV}$

THE BASICS OF WATER CHEMISTRY

Ordinary Liquid Water

Ordinary liquid water illustrates an important point: sometimes widely held, but not really examined, assumptions can turn out to be questionable. For example, what is the likely ionic configuration of a water molecule?

One possible ionic configuration for water is $2\,\text{H}^+ + \text{O}^{2-}$, and most people believe this is what ordinary water is. Let us now check that belief. The transition $\text{H} \rightarrow \text{H}^+$ takes:

$$IP_{1,1} / M_1 = 14.250 / 1.008 = 14.1369\text{eV} .$$

So 2H^+ takes $2 \times 14.1369 = 28.2738\text{eV}$. The transition $\text{O} \rightarrow \text{O}^{2-}$ takes:

$$-IP_{1,1}\left[\sqrt{Z(Z+1)}+\sqrt{Z(Z+2)}\right]/M_Z+\left[\Delta IP_{1,Z}Z-\Delta IP_{1,Z+2}(Z+2)\right]/M_Z$$
$$=-14.250\left[\sqrt{8\times 9}+\sqrt{8\times 10}\right]/15.999+\left[\Delta IP_{1,8}\times 8-\Delta IP_{1,10}\times 10\right]/15.999$$
$$=-14.250\left[\sqrt{72}+\sqrt{80}\right]/15.999+\left[13.031\times 8-29.391\times 10\right]/15.999$$
$$=-14.250\left[8.4853+8.9443\right]/15.999+\left[104.248-293.910\right]/15.999$$
$$=-14.250\left[17.4296\right]/15.999+\left[-189.662\right]/15.999$$
$$=-15.5242-11.8546=-27.3788\,\text{eV}\quad .$$

So the creation of a water molecule in the ionic configuration $2\,H^+ + O^{2-}$ takes altogether $28.2738-27.3788=0.8950\text{eV}$. This is a positive energy requirement, and it implies that some external assistance is needed. So normal water may *not* be in this ionic configuration after all!

But another possibility is readily at hand. The ionic configuration for normal water could be $2\,H^- + O^{2+}$. The transition $H \to H^-$ takes:

$$-IP_{1,1}\sqrt{Z(Z+1)}/M_Z-\Delta IP_{1,Z+1}(Z+1)/M_Z+\Delta IP_{1,Z}Z/M_Z$$
$$=-IP_{1,1}\sqrt{1\times 2}/M_1-\Delta IP_{1,2}\times 2/M_1+\Delta IP_{1,1}\times 1/M_1$$
$$=-14.250\times 1.4142/1.008-35.625\times 2/1.008+0\times 1/1.008$$
$$=-19.9924-70.6845+0=-90.6769\text{eV}.$$

So $2\,H^-$ takes $2\times(-90.6769)=-181.3538\text{eV}$. Please pause here and take notice: this is a *huge* negative energy. It reflects the fact that electrons *really* like to make pairs. They do it everywhere in Chemistry. Their propensity to do it motivated the invention of the so-called 'spin' quantum number. Without spin, many electrons in atoms would be violating the 'Pauli Exclusion Principle', which says that only one electron can be in any particular quantum state.

Electron pairs are famous in Condensed Matter Physics too, under the name 'Cooper pairs'.

Now proceeding, the transition $O \to O^{2+}$ takes:

$$IP_{1,1}\left[Z+\sqrt{Z(Z-1)}\right]/M_Z+\left[\Delta IP_{1,Z}Z-\Delta IP_{1,Z-2}(Z-2)\right]/M_Z$$
$$=IP_{1,1}\left[8+\sqrt{8\times 7}\right]/M_8+\left[\Delta IP_{1,8}\times 8-\Delta IP_{1,6}\times 6\right]/M_8$$
$$=14.250\left[8+\sqrt{56}\right]/15.999+\left[13.031\times 8-7.320\times 6\right]/15.999$$
$$=13.7907+3.7707=17.5614\,\text{eV}\,.$$

So the creation of the water molecule in the ionic configuration $2\,H^- + O^{2+}$ takes altogether $2\times(-90.6769)+17.5614=$ $-181.3538+17.5614=$ $-163.7924\,\text{eV}$. This is a solidly negative energy. It means that ordinary water is overwhelmingly in the second candidate ionic configuration, $2\,H^- + O^{2+}$. This *is* normal water.

However, in situations where more than one version of anything can exist, they generally all *do* exist, in proportions determined by their so-called Boltzmann factors, $\exp(-E/kT)$, where E is energy, k is Boltzmann's constant, and T is absolute temperature. Boltzmann factors are the evidence of entropy maximization at work. So ordinary liquid water will contain a tiny, tiny fraction of the first ionic configuration, $2H^+ + O^{2-}$.

EXCLUSION ZONE WATER

Prof. Gerald Pollack wrote the definitive book [15] on the subject of water in the newly discovered state characteristic of the surface of all water. This state is called Exclusion Zone Water, or EZ water. The name comes from the fact that EZ water tends to expel positive ions. It does this because it forms nested layers of hexagonal arrays of negative ions, each layer offset to the next, thus leaving no space for unused positive ions.

The following chemical transition is proposed for the creation of EZ water:

3 normal water molecules $\rightarrow$ 1 EZ water ion + 1 hydronium ion

That is, $3(H_2O)\rightarrow(H_3O_2)^- + (H_3O)^+$.

Observe the notation: parentheses are used to avoid any prejudice about what charge the individual atoms within any chemical species, whether neutral molecule or charged ion or radical, may carry. We want to consider all possible, or at least all plausible, ionic configurations of every chemical species mentioned.

The EZ water ion $(H_3O_2)^-$ could have the ionic configuration $3\,H^+ + 2\,O^{2-}$. The $3\,H^+$ takes total energy $3 \times IP_{1,1} / M_z =$ $3 \times 14.250 / 1.008 = 42.4107\,\text{eV}$. The $2\,O^{2-}$ takes total energy $2 \times (-27.3788) = -54.7576\,\text{eV}$. So the ionic configuration $3\,H^+ + 2\,O^{2-}$ takes $42.4107 - 54.7576 = -12.3469\,\text{eV}$. This is a negative energy, so this ionic configuration certainly can occur.

But there is another possibility too. The EZ water ion $(H_3O_2)^-$ could have the ionic configuration $3\,H^- + 2\,O^+$. An H^- takes $-90.6769\,\text{eV}$, so $3\,H^-$ takes $3 \times (-90.6769) = -272.0307\,\text{eV}$. An O^+ takes energy:

$$\begin{aligned}
&(IP_{1,1} \times 8 + \Delta IP_{1,8} \times 8 - \Delta IP_{1,7} \times 7) / M_8 \\
&= (14.250 \times 8 + 13.031 \times 8 - 13.031 \times 7) / 15.999 \\
&= (114.000 + 104.248 - 91.217) / 15.999 = 7.9399\,\text{eV}.
\end{aligned}$$

So $2\,O^+$ takes $2 \times 7.9399 = 15.8798\,\text{eV}$. Thus the ionic configuration $3\,H^- + 2\,O^+$ takes $-272.0307 + 15.8798 = -256.1509\,\text{eV}$. This energy is much more negative than that of the first candidate ionic configuration for EZ water, $3\,H^+ + 2\,O^{2-}$. This fact means EZ water is nearly always in this second candidate ionic configuration, $3\,H^- + 2\,O^+$.

As for the hydronium ion $(H_3O)^+$, it could have the ionic configuration $3\,H^+ + O^{2-}$. This, I believe, is what most people would expect, given that most people would have expected (wrongly, as it turned out) that normal water had the ionic configuration $2\,H^+ + O^{2-}$. From the study of regular water, we know the candidate $3\,H^+$ would take $3 \times 14.250 = 42.4107\,\text{eV}$, and the candidate O^{2-} would take $-27.3788\,\text{eV}$, so the candidate ionic configuration $3\,H^+ + O^{2-}$ for hydronium would take $42.4107 - 27.3788 = 15.0319\,\text{eV}$. This is a solidly positive energy, *and so the ionic configuration* $3\,\mathrm{H}^{+}+\mathrm{O}^{2-}$ *for the hydronium ion* $(\mathrm{H}_3\mathrm{O})^{+}$ *occurs only very infrequently.*

But, as it was with the EZ water ion, there is another possibility for the hydronium ion. The $(\mathrm{H}_3\mathrm{O})^{+}$ could have the ionic configuration $3\,\mathrm{H}^{-}+\mathrm{O}^{4+}$. The $3\,\mathrm{H}^{-}$ would take $3\times(-90.6769)=\ -272.0307\,\mathrm{eV}$, and the O^{4+} would take:

$$IP_{1,1}\left[Z+\sqrt{Z(Z-1)}+\sqrt{Z(Z-2)}+\sqrt{Z(Z-3)}\right]\Big/M_Z+\left[\Delta IP_{1,Z}Z-\Delta IP_{1,Z-4}(Z-4)\right]\Big/M_Z$$
$$=14.250\times\left[8+\sqrt{56}+\sqrt{48}+\sqrt{40}\right]\Big/15.999+\left[13.031\times8-9.077\times4\right]\Big/15.999$$
$$=14.250\times28.7361/15.999+67.940/15.999=25.5947+4.2465=29.8411\mathrm{eV}.$$

So the second candidate ionic configuration $3\,\mathrm{H}^{-}+\mathrm{O}^{4+}$ for the hydronium ion $(\mathrm{H}_3\mathrm{O})^{+}$ would take $-272.0307+29.8411=\ -242.1895\mathrm{eV}$. This is a robustly negative energy. It explains why the reaction product that accompanies EZ water *is* a hydronium ion, rather than a naked proton ($14.1369\mathrm{eV}$) next to a normal water molecule ($-163.7925\mathrm{eV}$), which would add up to $-149.6556\mathrm{eV}$, which, though negative, is not anywhere near as robustly so.

Together, the EZ water ion and the hydronium ion take $-268.6204-242.1895=\ -510.8099\mathrm{eV}$. Compare this to three normal water molecules at $3\times(-163.7925)=\ -491.3775\mathrm{eV}$. The EZ water ion together with the hydronium ion is favored over the three normal water molecules. That means nature stands poised to make EZ water ions and hydronium ions whenever it has the opportunity to do so. Generally speaking, the opportunity involves a surface and a little bit of incident energy, such as that provided by sunlight.

The other known states of water are Water Ice and Water Vapor, both of great interest to be sure, but probably not directly involved in CF.

The CF Application

The next question in the present study is this: what if any of the molecules or ions discussed here had been created with some heavy nuclei included?

That is, what happens to the ionic configurations if one Hydrogen atom H is replaced with a Deuterium D, or even a Tritium T?

where an H^+ ion takes:

$$IP_{1,1} \times 1 / M_H = IP_{1,1} \times 1/1 = 14.2500 \text{ eV},$$

a D^+ ion takes:

$$IP_{1,1} \times 1 / M_D = IP_{1,1} \times 1/2 = 14.250/2 = 7.1250 \text{ eV},$$

and a T^+ ion takes:

$$IP_{1,1} \times 1 / M_T = IP_{1,1} \times 1/3 = 14.250/3 = 4.7500 \text{ eV}.$$

The message is: heavier positive ions are easier to make.

Similarly, where an H^- ion takes:

$$-IP_{1,1}\sqrt{1\times 2}\big/M_H - \Delta IP_{1,2} \times 2 / M_H =$$
$$-14.250 \times 1.4142/1 - 35.6250 \times 2/1$$
$$= -20.1495 - 71.2500 = -91.3995 \text{ eV}$$

a D^- instead takes:

$$-IP_{1,1}\sqrt{1\times 2}\big/M_D - \Delta IP_{1,2} \times 2 / M_D =$$
$$-14.2500 \times 1.4142/2 - 35.6250 \times 2/2 =$$
$$-10.0762 - 35.6250 = -45.7012 \text{ eV}$$

and a T^- takes:

$$-IP_{1,1}\sqrt{1\times 2}\big/M_T - \Delta IP_{1,2} \times 2 / M_T =$$
$$-14.2500 \times 1.4142/3 - 35.6250 \times 2/3 =$$
$$-6.7175 - 23.7500 = -30.4675 \text{ eV}$$

The message is: heavier negative ions are less rewarding to make.

This is all to the good, inasmuch as CF requires naked nuclei, i.e., positive ions, as reactants. So can we get such reactants from water? In the case of normal water, we now know that the usual ionic configuration must be $2\,H^- + O^{2+}$, and not the previously presumed $2\,H^+ + O^{2-}$. The latter has a positive energy, and so does not occur abundantly.

But what about $D^+ + H^+ + O^{2-}$? We have above that D^+ takes 7.1250 eV, H^+ takes 14.2500 eV, and from the earlier analysis of ordinary water, we know O^{2-} takes -27.3788 eV. So $D^+ + H^+ + O^{2-}$ will take

$$7.1250 + 14.2500 - 27.3788 = -6.0038 \text{ eV} .$$

So in heavy water, this ionic configuration will occur naturally too.

In the case of EZ water, we know that, along with the more favored ionic configuration $3\,H^- + 2\,O^+$, the less favored $3\,H^+ + 2\,O^{2-}$ *can* exist naturally, and so probably *does* exist. So what about $D^+ + 2\,H^+ + 2\,O^{2-}$? That will take

$$7.1250 + 2 \times 14.2500 + 2 \times (-27.3788) =$$
$$7.1250 + 28.5000 - 54.7576 = -19.1326 \text{ eV} .$$

Compare this to $3\,H^+ + 2\,O^{2-}$ with pure Hydrogen, at

$$3 \times 14.250 / 1 - 54.7576 = 42.750 - 54.7576 = -12.0076 \text{ eV} .$$

The EZ ion with a Deuterium ion included is at *lower energy*, and so forms *more easily* than the one without Deuterium. This EZ ion will be preferred. So the action of forming EZ water can tend to concentrate the kind of reactant that CF requires.

But we should also look at the other ionic configuration for a deuterated EZ ion, $D^- + 2\,H^- + 2\,O^+$. The D^- takes

$$-IP_{1,1}\sqrt{Z(Z+1)}\big/M_Z - \Delta IP_{1,Z+1}(Z+1)/M_Z + \Delta IP_{1,Z} Z / M_Z$$
$$= -IP_{1,1}\sqrt{1 \times 2}\big/2 - \Delta IP_{1,2}\, 2/2 + \Delta IP_{1,1}/2$$
$$= -14.250 \times 1.4142 / 2 - 35.625 + 0 = -25.5486 \text{ eV}.$$

Above we found that pure H^- takes $-20.1495-71.2500=-91.3995$ eV, so $2\,H^-$ takes $2\times(-91.3995)=-182.7990$ eV. The $2\,O^+$ takes $2\times 7.9399=15.8798$ eV. So altogether, the ionic configuration $D^- + 2\,H^- + 2\,O^{2+}$ takes:

$$-25.5486-182.7990+15.8798=-192.4678 \text{ eV} .$$

This is a solidly negative energy, so this ionic configuration would certainly dominate the EZ – at least until such time a source of energy is introduced. That source would be the electrodes that are inserted to start the CF attempt. The process is analogous to the process for making Brown's gas: electrochemistry. In the CF case, when the electrodes are introduced they supply some energy, and the EZ ions can then go to their less favored, higher-energy, ionic configuration, wherein the deuterons are exposed, and CF can then ensue. This EZ story appears to play into a peculiar fact about CF that is presently known, but not well understood: CF generally goes better if the reactant heavy water is left to rest quietly for a long time before being used. This rest period was described as 'incubation' in a talk given by Senior VP Frank Gordon at Global Energy Corporation, Annandale, VA, speaking at a Conference of the Natural Philosophy Alliance in Long Beach, CA, in 2010.

Since then, I have come to think that three important processes may characterize 'incubation': 1) Gravity sends water molecules with heavier isotopes to lower levels in the water container, thus concentrating them more at locations lower in the water container; 2) The water container has internal surfaces, and they encourages EZ formation at those surfaces. There, the exclusion property of the EZ water further concentrates the positive ions that contain the heavy isotopes. 3) The stage is then set, but nothing dramatic can happen until electrodes are inserted, and a little energy is supplied to trigger the CF process. Where should the electrodes go? We may want to avoid, *or* to seek, the exclusion zones at the top, bottom, and sides of the container. We certainly want to reach positive ions of heavier mass, so we should put electrodes in the lower part of the water container. Should the current go horizontal or vertical, and if vertical, which way? We need experimental investigation of all these questions.

Conclusion

CF is an Engineering Problem, and it demands Engineering Analyses. This paper offers some suggestions as to what the engineering analyses should be like.

1) Engineering is somewhat like Applied Mathematics. The DSB protocol is to Applied Mathematics what First Aid is to Medicine: Step 1. In many cases, it solves the whole problem at hand. Only when it does *not* seem to solve the whole problem should the practitioner turn to anything else. So I believe the Engineering Analysis of CF should definitely use the DSB Protocol.
2) Engineering is usually *not* like Pure Mathematics: engineers do not generally involve Postulates in Engineering Analyses. What they do instead is list their Assumptions. The word 'Assumption' is much milder than the word 'Postulate'; it allows for a possible update, such as may occur in a typical engineering design review. So I believe engineers should not feel bound by any pre-established Postulate, such as the Second Postulate of SRT.

I am today hopeful that we can solve many of the engineering problems presented by CF. In so doing, we might also help eliminate the many Paradoxes that come into Physics along with the Second Postulate of SRT. The present situation is *itself* quite paradoxical: It is generally considered *radical* to do the actually *traditional* thing; i.e., apply the DSB Protocol, and it is generally considered *traditional* to do the originally *radical* thing; i.e., embrace the Second Postulate!

References

[1] C.K. Whitney, *Algebraic Chemistry, Applications and Origins*, Nova Science Publishers, Hauppauge, NY, USA (2013).

[2] A. Einstein, "On a Heuristic Point of View Concerning the Production and Transformation of Light", Annalen der Physik 17, 132-148, 1905, reprinted in *The Collected Papers of Albert Einstein*, Vol. 2, pp. 86 – 103, Princeton University Press, (1989).

[3] A. Einstein, "On the Electrodynamics of Moving Bodies", Annalen der Physik 17, 891-921, 1905, reprinted in *The Collected Papers of Albert Einstein*, Vol. 2, pp. 140 – 173, Princeton University Press (1989).

[4] R.P. Feynman, QED, *The Strange Theory of Light and Matter*, Princeton University Press, Princeton, NJ (1985).

[5] L. Brilloin, *Science and Information Theory*, Second Ed., Dover Publications, Mineola, NY (1962).

[6] http://www.britannica.com/EBchecked/topic/194901/Euclidean-geometry.

[7] J.D. Jackson, *Classical Electrodynamics*, Sect. 14.1, Second Ed., John Wiley & Sons, New York, (1962, 1975).

[8] W. Ritz, *Recherches critiques sur l'Électrodynamique Générale*, *Annales de Chimie et de Physique* 13: 145–275 (1908).

[9] V. Kuligin, M. Korneva, G. Kuligina, "Lorentz Transformation and Image Distortion", Galilean Electrodynamics 26, Special Issue 2, Fall (2015) (forthcoming).

[10] A. Liènard, "Champ Electrique et Magnètique produit par une Charge Electrique Concentrée en un Point et Animée d'un Movement Quelconque", *L'Eclairage Electrique vol. XVI* (27) 5-14, (28) 53-59, (29) 106-112 (1898).

[11] E. Wiechert, "Elektrodynamische Elementargesetze", *Archives Néerlandesises des Sciences Exactes et Naturelles série II, Tome IV* (1901).

[12] L.H. Thomas, "The Kinematics of an Electron with an Axis", Phil. Mag. Series 7, 3 (13) 1 – 22 (1927).

[13] J.P. Wesley, *Classical Quantum Theory*, Capter 7, Benjamin Wesley, 78176 Blumberg, Germany, 1996.

[14] C. Whitney, "EZ Water: A New Quantitative Approach Applied – Many Numerical Results Obtained", Water 5, 105-120 (2013).

[15] G. Pollack, *The Fourth Phase of Water: Beyond Solid, Liquid, and Vapor*, Ebner & Sons, Seattle WA, USA (2013).

In: Fusion Energy and Power
Editor: Lionel Romero
ISBN: 978-1-63482-548-1

Chapter 4

CONTROL CONCEPT FOR THE HIGH DENSITY AND LOW TEMPERATURE IGNITION IN THE FFHR HELICAL REACTOR

O. Mitarai[1,*], A. Sagara[2] and R. Sakamoto[2]
[1]Kumamoto Liberal Arts Education Center, Tokai University, Toroku, Higashi-ku, Kumamoto, Japan
[2]National Institute for Fusion Science, 322-6 Oroshi-cho, Toki, Japan

ABSTRACT

Super high dense core (SDC) plasma with more than $1x10^{21}$ m^{-3} observed in the large helical device (LHD) stimulates the creation of concept of controlling the unstable operating point in the high density and low temperature regime for the Force Free Helical Reactor (FFHR). This novel control method discovered in 2006 can stabilize the unstable operating point using proportional- integral-derivative (PID) control of the fueling with the error of the fusion power of $e(P_f)= -(P_{fo}-P_f)$. Here $P_{fo}(t)$ is the fusion power set value and $P_f(t)$ is the measured fusion power. Although continuous fueling with the small density variation has been actively studied at the unstable operating point, it is still uncertain whether the operating point can be stabilized when discrete fueling such as pellet injections is used.

* Corresponding author: E-mail: omitarai@ktmail.tokai-u.jp.

In this work, we show at first that it is possible to control the unstable ignited operation for the density variations corresponding to the pellet size of 12 mm~16 mm in FFHR2m (R=14 m, $\bar{a}$ =1.73 m). In addition, in the commercial reactor FFHR2m2 (R=15.7 m, $\bar{a}$ =2.5 m,) the external heating power is specified to be the foreseeable neutral beam injection (NBI) system with the energy of 1.5 MeV. In FFHR2m2 the initial density is kept at the low value less than ~$4x10^{20}$ m^{-3} to overpass the saddle point on Plasma Operation Contour (POPCON) map by the neutral beam injection heating. After passing over the saddle point, a high-density ignition is accessed by alpha heating after turning off NBI injection. Here the control algorithm of the NBI external heating power has been newly implemented based on the temperature limit concept instead of the preprogramming used in FFHR2m. Various disturbances in plasma parameters due to impurity injection and confinement factor change can also be stabilized to some extent at the thermally unstable point by the feedback control of the heating power and discrete fueling.

1. INTRODUCTION

Remarkable result of the super dense core (SDC) plasma up to $1.1x10^{21}$ m^{-3} has been obtained in recent large helical device (LHD) pellet injection experiments [1][2]. In Wendelstein 7-AS, the high density H-mode with $4x10^{20}$ m^{-3} has also been produced [3]. Such low temperature and high density operation is generally advantageous to reduce the divertor heat flux in both tokamak and helical reactors. However, an operation in the low temperature and high-density ignition regime usually suffers from the thermal instability, where the operating point moves to the higher temperature side in the ignition regime or lower temperature side in the sub-ignition regime. New control algorithm using the simple PID control on fueling has been proposed to stabilize the thermal instability in the LHD type helical reactor (FFHR2m) [4]-[6]. Owing to the simple and comprehensive algorithm which does not need any linearization of the particle and power balance equations, this control method can be implemented in a reactor where the fusion power measurement provides the feedback signals for PID control of fueling.

In 1969 Mills pointed out that a thermonuclear reactor could be thermally unstable [7]. Since then, the control method of the unstable ignition in a fusion reactor had been studied [8]. Many stabilizing methods, such as by the fueling [9], by the compression-decompression of the plasma using the vertical field [10]-[12], by controlling alpha heating [13], and by toroidal field ripples [14], had been proposed in early phase of the nuclear fusion research. After 1985, also many stabilizing methods of the thermally unstable ignition have been proposed by such as fueling [15], impurity injection [16] and heating power modulations [17], and these combinations. In these controls, zero-dimensional equations of the particle and power balance equations have been linearized around the unstable operating point, and then stabilizing technique such as H-infinity control is applied [18]. Recently, the model based non-linear control method has been developed to avoid linearization in many equations [19], and non-model based control technique such as neural network control has been developed [20]. Among these studies the references [16] and [17] pointed out many advantages of the unstable low temperature and high density operation regime, because the divertor heat load can be reduced due to the strong Bremsstrahlung radiation, the high recycling and low divertor temperature operation is possible to minimize the sputtering at the divertor plate, and more efficient fueling is possible by deep penetration of pellets.

However, in above methods, linearization is necessary in many equations or many plasma parameters should be measured or estimated to apply the heating power and fueling in the nonlinear control method [19]. Hence it may be difficult to apply it to the actual situation where the limited numbers of plasma diagnostics are only available. In those works only stabilization around the steady state unstable operating point has been shown, but the transient phase to the unstable operating point from the initial low temperature and low density has not been demonstrated although such possibility was described. Although it is discussed in those works, it is not certain whether the model based control method can be used in actual experimental situations when the profile changes. On the other hand, the new PID control method recently proposed by authors does not require any linearization, and the fusion neutron measurement is basically enough to control the thermally unstable operating point [4]-[6]. We demonstrated that by switching the polarity of the coefficient in the error of the fusion power the operating point can move to the unstable operating point from the stable operating point, and the ignition access to the thermally unstable ignition boundary is possible from the initial low temperature and density state. As this PID control is synthesized without any dynamic model of the system, which makes it inherently robust against model uncertainties, it can be applied to 1-D simulation code and implemented in a fusion reactor. As this controller uses only the total fusion power, it can be used when the density and temperature profiles change. As fueling was continuous in previous works [4]-[6], further works on the control of the thermally unstable operation are needed for discrete fueling such as pellet injections.

In this study, fueling is discontinuously provided to simulate pellet injections in SDC discharge instead of continuous fueling such as gas puffing. Despite of parameter variations due to discrete fueling, we have found that it is possible to control the thermally unstable operating point in FFHR2m. However when the density variation is too large, the ignition cannot be maintained. As the heating system is not specified and is controlled by preprogramming in FFHR2m, the feedback controlled NBI is newly employed as a main heating system in FFHR2m2 [21]. Here the initial density is kept at the low value less than ~4×10^{20} m^{-3} to overpass the saddle point on POPCON by NBI heating. After passing the saddle point, a high-density ignition is accessed by alpha heating after turning off NBI injection. This control algorithm of the NBI heating power has been invented based on the temperature limit concept [22]. We have also found that this stabilizing control can be used to some extents for various disturbances such as impurity increment and the change in the confinement factor [23]. Thus, we believe that the unresolved issued on the control of thermally unstable operation pointed out at 1969 [7] has been solved by authors.

In section 2, zero-dimensional equations, control algorithm of the continuous and discrete fueling for the thermally unstable operation are described. Discrete fueling corresponding to pellet injection is described with pellet size. In section 3, temporal evolutions of ignition access to the unstable operating point by continuous fueling, and discrete fueling are presented. In section 4, the external heating method is specified to neutral beam heating injection (NBI), the feedback control algorithm of the heating power and discrete fueling effect on the thermally unstable ignition are presented. In section 5 was shown the control robustness against the parameter changes at the unstable operating point. In section 6, discussion and summary are given. In Appendix-1 and -2, particle and power balance equations, and pellet penetration depth and its dependency on the temperature and density profiles are described in detail, respectively.

2. Formalism and Control Algorithm

2.1. Zero-Dimensional Equations

In this analysis, the global (combined) particle balance equation, alpha ash particle equation and the power balance equation are used [5]. The detailed formulas are described in Appendix-1.

(a) Combined particle balance equation

The density evolution is calculated by the combined particle balance equation, which is given by the deuterium, tritium, and alpha particle balance equations together with the charge neutrality condition

$$\frac{dn_e(0)}{dt} = \frac{1}{1-8f_O}\left[(1+\alpha_n)S_{DT}(t) - \left\{\frac{f_D+f_T}{\tau_p^*} + \frac{2f_\alpha}{\tau_\alpha^*}\right\}n_e(0)\right] \tag{2-1}$$

where f_o is the oxygen impurity fraction, a_n is the density profile factor, S_{DT} is the D-T fueling rate given by the continuous or discrete fueling in this study, f_D is the deuterium fraction, f_T is the tritium fraction, f_α is the alpha ash fraction, τ_p* is the effective D-T fuel particle confinement time, and τ_α* is the effective alpha ash confinement time. (Note the electron density obtained from Eq. (2-1) is plotted in the column (a) in the following figures.)

(b) Alpha ash particle balance equation

The temporal evolution of alpha ash fraction is calculated by

$$\frac{df_\alpha}{dt} = (1+\alpha_n)n_e(0)f_D f_T \overline{\langle\sigma V\rangle_{DT}(x)} - \frac{f_\alpha}{\tau_\alpha^*} - \frac{f_\alpha}{n_e(0)}\left[\frac{dn_e(0)}{dt}\right] \tag{2-2}$$

Here the fuel ratio of D:T=50:50 is assumed and hence $f_D=f_T$, $f_D = (1- 2f_\alpha - 8f_o)/2$, and $Z_{eff} = 2f_D + 4\ f_\alpha + 64\ f_o$. The alpha ash confinement time ratio of $\tau_\alpha^*/\tau_E=3$ or 4 and the fuel particle confinement time ratio of $\tau_p^*/\tau_E=3$ have been used as a reference in Eqs. (2-1) and (2-2). (Note the calculated alpha ash fraction by Eq. (2-2) is plotted in the box (b) in the following figures.) As alpha ash confinement time ratio τ_α^*/τ_E plays an important role in a high-density operation, such dependence was studied in detail in the previous paper [5][6]. We note that the alpha ash confinement time ratio is simply estimated as τ_α^*/τ_E =5.9~9 in LHD experiments using He gas puffing method [6], although further careful study is necessary. In that paper, wide parameter regime with τ_α^*/τ_E=3 ~8 was surveyed and we have used optimistic values of τ_α^*/τ_E=3 as a reference to expand the ignition regime for studying ignition characteristics with various disturbances. The fuel particle confinement time ratio τ_p^*/τ_E is also important parameter in the unstable operation as discussed in the previous paper [4][5]. We have surveyed ignition characteristics up to

$\tau_p^*/\tau_E \sim 7$ in that work, but we use $\tau_p^*/\tau_E = 3$ as a reference unless otherwise noted.

(c) Power balance equation

The power balance equation is

$$\frac{d\overline{W}}{dt} = P_{EXT} / V_o - \left(\overline{P}_L + \overline{P}_B + \overline{P}_S - \overline{P}_\alpha\right) = P_{NET} - \overline{P}_L \tag{2-3}$$

where P_{EXT} is the external heating power, V_o is the plasma volume, $\overline{P}_L$ is the plasma conduction loss per unit volume, $\overline{P}_B$ is the Bremsstrahlung loss power density, $\overline{P}_S$ is the synchrotron radiation loss power per unit volume which is negligible in the low temperature operation, and $\overline{P}_\alpha$ is the alpha heating power density. The net heating power density, $P_{NET} = P_{EXT} / V_o + \overline{P}_\alpha - \overline{P}_B - \overline{P}_S$, is used for calculating the plasma conduction loss (Eq. (A-6)) and the density limit (Eq. (A-9)).

The ion temperature evolution is calculated by the power balance equation after volume averaging as

$$\begin{aligned} \frac{dT_i(0)}{dt} &= \frac{1+\alpha_n+\alpha_T}{1.5e\left(f_D+f_T+1/\gamma_i+f_\alpha\right)n_e(0)} \times \left[\left\{P_{EXT}/V_o+\overline{P}_\alpha\right\} - \left\{\overline{P}_L+\overline{P}_b+\overline{P}_s\right\}\right] \\ &- \frac{T_i(0)}{\left(f_D+f_T+1/\gamma_i+f_\alpha\right)} \times \left[\left\{1-8f_O+\frac{1}{\gamma_i}-f_\alpha\right\}\frac{1}{n_e(0)}\left[\frac{dn_e(0)}{dt}\right] - \left[\frac{df_\alpha(0)}{dt}\right]\right] \end{aligned} \tag{2-4}$$

where equal ion and electron temperature $\gamma_i = T_i(0)/T_e(0) = 1$ is assumed. Detailed derivation is described in the Appendix-1. (Note the calculated temperature is plotted in the box (a) in the following figures.)

Above Eqs. (2-1), (2-2) and (2-4) are solved simultaneously for the density, alpha ash fraction and ion temperature evolutions using various power loss terms estimated for the box type and parabolic density profile as given in section 2.2.

(d) Confinement scaling law

Here, the following ISS95 confinement scaling [24] has been used for estimation of the plasma conduction loss.

$$\begin{cases} \tau_E[s] = \gamma_{ISS}\ \tau_{ISS}[s] = \gamma_{LHD} \times 1.6\ \tau_{ISS}[s] \\ \tau_{ISS}[s] = 0.079\,\iota_{2/3}^{0.4}\,\bar{n}_{19}^{\;0.51}[10^{19} m^{-3}] B_o^{0.83}[T] \bar{a}^{2.21}[m] R^{0.65}[m] / P_{HT}^{0.59}[MW] \end{cases} \tag{2-5}$$

where γ_{ISS} and γ_{LHD} represent the confinement enhancement factors over the ISS95 and present LHD scalings, respectively, and $\iota_{2/3}$ is the rotational transform at the 2/3 of the normalized minor radius, $\bar{n}_{19}$ is the line density, B_o is the magnetic field strength, $\bar{a}$ is the effective minor radius, R is the major radius, and P_{HT} is the heating power. γ_{ISS}=1.6 (γ_{LHD}= 1.0) and 1.43 (γ_{LHD}= 0.89) have been used for FFHR2m and FFHR2m2 in this study, respectively. The detailed procedure of the plasma conduction loss $\bar{P}_L$ is described in the Appendix-1.

Sudo density limit scaling $n(0)_{lim}$, Eq. (A-9) in Appendix-1, on the line density of the core plasma with the density limit factor of γ_{SUDO}=5.5 is used as a measure of density [25]. In the following figures, $n(0)_{lim}$ is plotted in the box (a). In actual LHD experiments, as the density limit is determined by the edge density and temperature [26], the high density is not critical as long as the edge density is lower than the Sudo density limit converted to the edge density. The external heating power P_{EXT} is preprogrammed in FFHR2m, and is controlled by feedback in FFHR2m2.

A birds-eye view of FFHR reactor is shown in Figure 1 and its machine parameters [28] are listed in Table 1.

Figure 1. Birds-eye view of FFHR2m helical reactor.

Table 1. Plasma parameters at the unstable and stable operating point in FFHR reactors

		FFHR2m	**FFHR2m2**	
		Unstable operating point	**Stable operating point**	**Unstable operating point**
		SDC profile	**Parabolic profile**	
Major radius	R (m)	14.0	15.7	
Effective minor radius	$\bar{a}$ (m)	1.73	2.5	
Plasma Volume	V_o (m^3)	827	1936	
Polarity/Field period	ℓ /m	2/10		
Coil pitch parameter	γ	1.15	1.2	
Magnetic filed	B_o (T)	6.0	4.5	
Maximum magnetic field	B_{omax} (T)	13.3	12.1	
Coil magnetic energy	W_c(GJ)	120	160	
Blanket thickness	ΔB(m)	1.2		
Rotational transform	$\bar{\iota}_{2/3}$	0.92		
Maximum external heating power	P_{EXT} (MW)	100		
Confinement factor over ISS95 scaling	γ_{ISS}	1.6	1.92	1.43
Confinement time	τ_E (s)	4.7	2.3	4.3
Alpha ash fraction	f_α	0.058	0.041	
Oxygen impurity fraction	f_o	0.0075		
Effective ion charge	Z_{eff}	1.48	1.50	
He ash confinement time ratio	τ_α^*/τ_E	3	4	
Fuel particle confinement time ratio	τ_p^*/τ_E	3		
Fusion alpha heating efficiency	η_α	0.9	0.98	
Operation density	$n_e(0)$ (10^{20} m^{-3})	9.6	1.8	8.4
Density limit factor over Sudo scaling	γ_{SUDO}	5.5	1.5	6.5
Density limit margin in the steady state	$[n(0)_{limit}/n(0)]$	1.59	1.24	1.41
Ion temperature	$T_i(0)$ (keV)	6.4	17.8	7.1
Ion to electron temperature ratio	T_i/T_e	1.0		

		FFHR2m	FFHR2m2	
		Unstable operating point	**Stable operating point**	**Unstable operating point**
		SDC profile	**Parabolic profile**	
Density profile	α_n	Box type	0.5	3.0
Temperature profile	α_T	0.25	1.0	
Beta value	$<\beta>$(%)	2.5	4.9	4.5
Total fusion power	P_f (MW)	1900	3000	
Total neutron power	P_n (MW)	1520	2400	
Total alpha heating power	P_α (MW)	380x0.9	600x0.98=588	
Total Bremsstrahlung power	P_B (MW)	248	87	348
Total synchrotron radiation power	P_S (MW)	0	4.2	0
Total plasma conduction loss	P_L (MW)	96	497	240
Electric power output (thermal efficiency)	P_e (MW)	627 (33%)	1000 (33%)	
Neutron wall loading	Γ_n (MW/m^2)	1.5	1.48	
Heat flux to first wall	Γ_h (MW/m^2)	0.25	0.056	0.22
Heat flux to divertor for 0.1 m wet width	Γ_{div}(MW/m^2)	5.4	25.0	12.2

2.2. Profiles for SDC Plasma and Corresponding Power Terms

When temperature and density profiles are given by the parabolic profile, the various loss terms and the plasma energy in the power balance equation can be calculated simply. However, in SDC discharge their profiles are quite different, and the various loss power profiles including the radiation power profile are changed. Therefore, in 0-dimensional analysis the volume averaged power losses needs to be reevaluated. In FFHR2m study we simply assumed the box type density profile as shown in Figure 2 to simulate SDC plasma using hyperbolic tangent and used the broad temperature profile with α_T =0.25 as observed in LHD experiments [1]. The coefficients of the various power loss terms are calculated for this profile. Here, the box type density profile is defined by

$$n(x)/n(0) = 0.5\left[1 - Tanh\left\{20\left(x^2 - 0.2\right)\right\}\right] \quad (2\text{-}6)$$

where Tanh is the hyperbolic tangent, and the temperature profile is

$$T(x)/T(0)=\left(1-x^2\right)^{\alpha_T} \tag{2-7}$$

For the parabolic density profile $n(x)/n(0)=(1-x^2)^{\rho_n}$ and the temperature profile (Eq. 2-7)), the coefficient of volume averaged plasma energy is defined by

$$\bar{W}/3kn(0)T(0)=\int_0^1 (1\text{-}x^2)^{\alpha_n}(1\text{-}x^2)^{\alpha_T}\, 2x\,dx=\frac{1}{1+\alpha_n+\alpha_T} \tag{2-8}$$

This coefficient is reevaluated for the box type density profile.

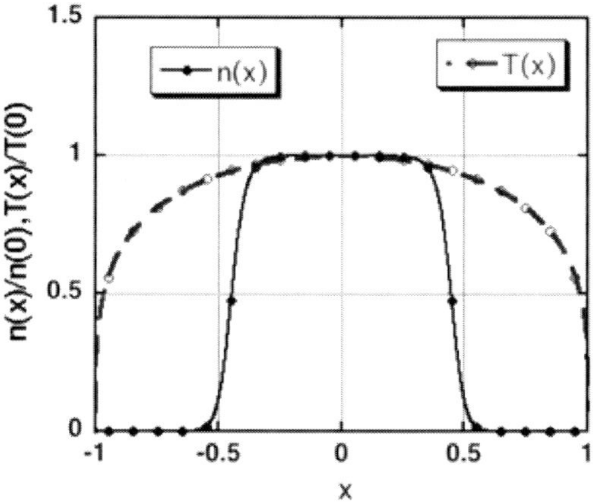

Figure 2. Box type density and parabolic temperature profiles for SDC discharge.

(a) Plasma energy

As the coefficient of volume averaged plasma energy per unit volume for SDC profile, $1/(1+\alpha_n+\alpha_T)$, is replaced by

$$\bar{W}/3kn(0)T(0)=\int_0^1 0.5\left[1-Tanh\left\{20\left(x^2-0.2\right)\right\}\right]\times\left(1-x^2\right)^{0.25}\, 2x\,dx=0.194429$$

Therefore, the final plasma energy loss is given by

$$\overline{W} = 0.194429\{3kn(0)T(0)\} \tag{2-9}$$

In the ion temperature equation (Eq. 2-4) for the box type density profile, the coefficient of volume averaged plasma energy should be replaced by $1/(1+\alpha_n+\alpha_T) = 0.194429$.

(b) Plasma Conduction Loss Power Per Unit Volume

The plasma conduction loss power per unit volume $\bar{p}_L$ is obtained from Eq. (A-7) in Appendix-1 as

$$\overline{P}_L = \left[\frac{A_L n(0)}{C_{op}'}\right]^{\frac{1}{1-S}} \left(10^{-6}\right)^{\frac{S}{1-S}} \tag{2-10}$$

where S=0.59 is the power of the heating power in the ISS95 scaling law, and

$$A_L = 1.5e\{f_D + f_T + 1/\gamma_i + f_\alpha\}T_i(0)/(1+\alpha_n+\alpha_T)$$

$$C_{op}' = \gamma_{ISS} \times 0.079 \iota_{2/3}^{0.4} \overline{n}_{19}^{0.51}[\times 10^{19} m^{-3}] B_o^{0.83}[T] \overline{a}^{2.21} R^{0.65}[m] / \left(2\pi R \pi \overline{a}^2\right)^S$$

Here the coefficient $1/(1+\alpha_n+\alpha_T)=0.194429$ should be used.

(c) Bremsstrahlung loss power density

As the coefficient of volume averaged Bremsstrahlung loss power density is replaced by

$$\frac{1}{1+2\alpha_n+0.5\alpha_T} = \overline{P}_B / \left\{1.488\times 10^{-38} Z_{eff} n(0)^2 \sqrt{T(0)}\right\}$$
$$= \int_0^1 \left\{0.5\left[1 - Tanh\left\{20\left(x^2-0.2\right)\right\}\right]\right\}^2 \times \left(1-x^2\right)^{0.25/2} 2x\,dx = 0.172891$$

the final Bremsstrahlung loss power density is given by

$$\overline{P}_B = 0.172891\left\{1.488\times 10^{-38} Z_{eff} n(0)^2 \sqrt{T(0)}\right\} \tag{2-11}$$

(d) Alpha heating power density

The volume-averaged alpha heating power density $\overline{P}_\alpha$ is directly integrated.

(e) Line-averaged density

The coefficient of line-averaged density is given by

$$\overline{n}/n(0) = \int_0^1 0.5\left[1 - Tanh\left\{20\left(x^2 - 0.2\right)\right\}\right]dx = 0.444073 \tag{2-12}$$

which is used in the ISS95 confinement time scaling.

(f) Volume averaged density

The coefficient of the volume-averaged density is given by

$$\frac{1}{1+\alpha_n} = \langle n\rangle / n(0) = \int_0^1 0.5\left[1 - Tanh\left\{20\left(x^2 - 0.2\right)\right\}\right]2x\,dx = 0.20 \tag{2-13}$$

Therefore, $1+\alpha_n=5$ is used for estimating the volume-averaged density in Eq. (2-1) and the volume averaged alpha ash fraction in Eq. (2-2).

Above relations (2-9) to (2-13) have been used in the power and particle balance equations for the box type or parabolic density and broad temperature profiles.

2.3. Unstable Ignition Control Algorithm and Pellet Size

2.3.1. Algorithm for continuous fueling

Stable ignition in the FFHR reactor is controlled by the PID continuous D-T fueling rate [27]:

$$S_{DT}(t) = S_{DT0}\left\{e_{DT}(P_f) + \frac{1}{T_{\text{int}}}\int_0^t e_{DT}(P_f)dt + T_d\frac{de_{DT}(P_f)}{dt}\right\}G_{fo}(t) \tag{2-14}$$

where $S_{DT0}=4\text{x}10^{19}$ m^{-3}/s, T_{int} is the integral time, T_d is the derivative time, and the gain $G_{fo}(t)$ increases from 2 to 50, and the error of the fusion power is

$$e_{DT}\left(P_f\right)=+\left(1-P_f(t)/P_{f0}(t)\right), \tag{2-15}$$

where $P_{fo}(t)$ is the fusion power set value and $P_f(t)$ is the measured fusion power [28]. However, in the unstable regime, the PID continuous D-T fueling rate, $S_{DT}(t)$, based on the error of the fusion power with a negative sign

$$e_{DT}\left(P_f\right)=-\left(1-P_f(t)/P_{f0}(t)\right) \tag{2-16}$$

can stabilize the thermal instability [4]-[6].

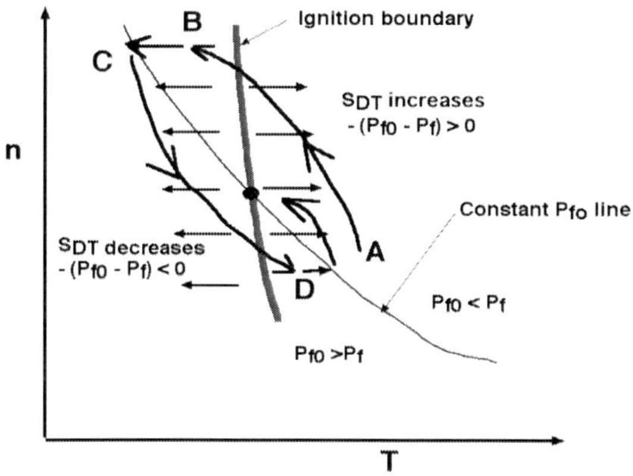

Figure 3. Schematic movement of the operating point around the unstable ignition boundary on POPCON.

This stabilizing mechanism is understood as shown in POPCON in Figure 3 for the continuous fueling. When P_f is larger than P_{fo}, the operating point (A) moves toward the higher density and lower temperature side. This operating point slightly shifts to the higher temperature side due to ignition nature between (A) and (B). When it enters in the sub-ignition regime (B), it goes to the lower temperature side due to sub-ignition nature and crosses the constant P_{fo} line (C). The fueling is now decreased and the operating point proceeds to the lower density and higher temperature side, and goes into the ignition regime (D), and crosses the constant P_{fo} line. Thus, oscillations take place and are damped away.

2.3.2. Algorithm for discrete fueling

Discrete fueling such as pellet injection is determined by the following PID signal, Error(P_f), based on the error of the fusion power, $e_{DT}\left(P_f\right) = -\left(1 - P_f(t)/P_{f0}(t)\right)$, as

$$Error(P_f) = \left\{ e_{DT}(P_f) + \frac{1}{T_{\text{int}}}\int_0^t e_{DT}(P_f)dt + T_d \frac{de_{DT}(P_f)}{dt} \right\} \tag{2-17}$$

where the fueling gain is set to unity due to digitizing. Therefore, no gain control exists in this discrete fueling.

The following control algorithm regulates discrete fueling:

$$\begin{cases} S_{DT}(t) = S_{DTpell} & \textit{for } Error(P_f) > 0 \\ S_{DT}(t) = 0 & \textit{for } Error(P_f) \le 0 \end{cases} \tag{2-18}$$

where S_{DTpell} is the fuel particle density provided by the one discrete fueling (one pellet train) as given belows.

2.3.3. Parameters for discrete fueling such as pellet size

One candidate of the discrete fueling is the repetitive pellet injections. Although the penetration problem exists for the case of pellet injection, we assume the full penetration of the pellet in this 0-dimensional study. Based on the time step of Δt=20 ms in the calculation, fueling is supplied at the three consecutive times (1 pellet train), and no fueling during the next three time steps as shown in Figure 4-(b). For the pellet speed of 1000 m/s, the transportation time of the pellet from the plasma edge to the plasma center (~1.2 m) is almost 1 ms in the FFHR2m. This transportation time is shorter than the time step of Δt=20 ms. Here all the injected pellets are assumed to be evaporated immediately and completely absorbed in the plasma. Therefore, the minimum repetition time of the pellet injection is T_{rep}=120 ms. (We note that the total transportation time is ~30 ms for the distance of ~30 m from pellet injector to the plasma. But this time delay is not taken into account in this study.) Detailed pellet injection algorithm was described in the reference [29]. The total fuelled particle number for continuous fueling and discrete fueling should be the same in the steady state. This fact would be checked in the section 3.2.

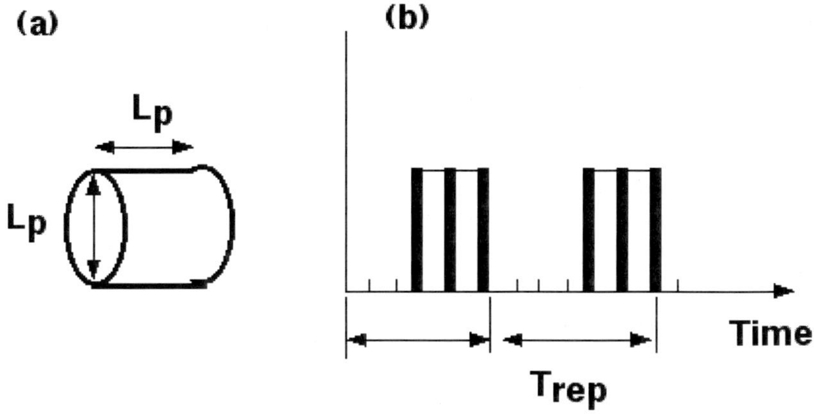

Figure 4. (a) Pellet dimension and (b) three pellet pulses (pellet train) based on the time step of Δt=20 ms. (T_{rep}=120 ms)

As the D-T solid molar volume is 19.88 mm^3/mol [30], the D-T ice density is given by {$6.02x10^{23}$x2}/19.88 [mm^3/mol]=$6.05x10^{28}$ m^{-3}. For the pellet size of L_p=12 mm diameter and L_p=12 mm length as shown in Figure 4-(a), the total D-T particle number is $N_{pell}=\pi(L_p/2)^2L_px6.05x10^{28} = 82x10^{21}$ m^{-3}. Therefore, the fueling particle number density divided by the plasma volume V_o is $S_{DTpell}=N_{pell}/V_o$= $0.99x10^{20}$ m^{-3} corresponding to the fueling rate in continuous fueling (see Figure 7-(d)). For the larger pellet size of 14 mm and 16 mm, the fueling particle number densities are S_{DTpell}= $1.57x10^{20}$ m^{-3} (see Figure 9-(d)) and S_{DTpell}=$2.35x10^{20}$ m^{-3} (see Figure 11-(d)), respectively.

The pellet penetration depth has been calculated based on the neutral gas-shielding (NGS) model [31], which is described in detail in Appendix-2, for two types of the density and temperature profiles. For both cases the pellet speed is assumed to be 1000 m/s. The first "SDC" profile (Figure 2) was employed in this work because the pellet penetration depth or the radial position of the source term does not depend on the temperature. As the pellet deposition is located inside a half of the normalized minor radius as shown in Figure A-1, the assumed "SDC" density profile could be produced. As the pellet deposition positions are not much different for 12 mm, 14 mm and 16 mm pellet sizes, the same density profile is used for simplicity. However, the local perturbation effect on the profile should be taken into account in future 0-dimensional analysis.

For further study of the pellet penetration, the different density and temperature profiles were also calculated as shown in Figure A-2. The pellet penetration depth now depends on the temperature. Above the temperature of 8 keV, the pellet penetration depth does not change appreciably. Thus we have found that the low temperature is suitable for better pellet penetration. We note that the temperature of 6 to 7 keV is obtained in this FFHR study.

3. Ignition Access to the Unstable Operating Point in FFHR2m

3.1. Continuous Fueling Case

For the SDC discharge, the unstable ignited operating point is directly accessed, and the fusion power rise-up time is taken as shorter, similar to the LHD experimental results [2], than the case of parabolic profiles described in the previous paper [4][5]. During the unstable ignition access the density is initially raised up to $\sim 6\text{x}10^{20}$ m^{-3} in the short time from 10 s to 12.8 s as schematically shown in Figure 5. The peak density is controlled by the error of the peak density as given by

$$e_{DT}(n) = +\left(1 - n(t)/n_{set}(t)\right) \qquad (3\text{-}1)$$

where $n_{set}(t)$ is the set value of the peak density. Initial set value of the peak density determines the density rise-up. Here, preprogrammed values are $n_{set}=0.05\text{x}10^{21}$ m^{-3} at t=10 s and $n_{set}=0.6\text{x}10^{21}$ m^{-3} at t=13 s. At 12.8 s, the control is switched to the fusion power control with $e_{DT}(P_f) = -\left(1 - P_f(t)/P_{f0}(t)\right)$ till the end of the discharge. Smooth transition from the density control to the fusion power control is important in this operation scenario, which was adjusted carefully.

Although continuous fuelling discharge is different from the actual SDC discharge where pellet injection is essential, it provides the basic operation scenario for discrete fueling. However, as the pellet injection is discrete fueling, fueling is delayed at least by the pellet repetition time. Although the pellet injection may alter the waveforms of the parameter evolution by discrete nature, it should essentially provide the same result.

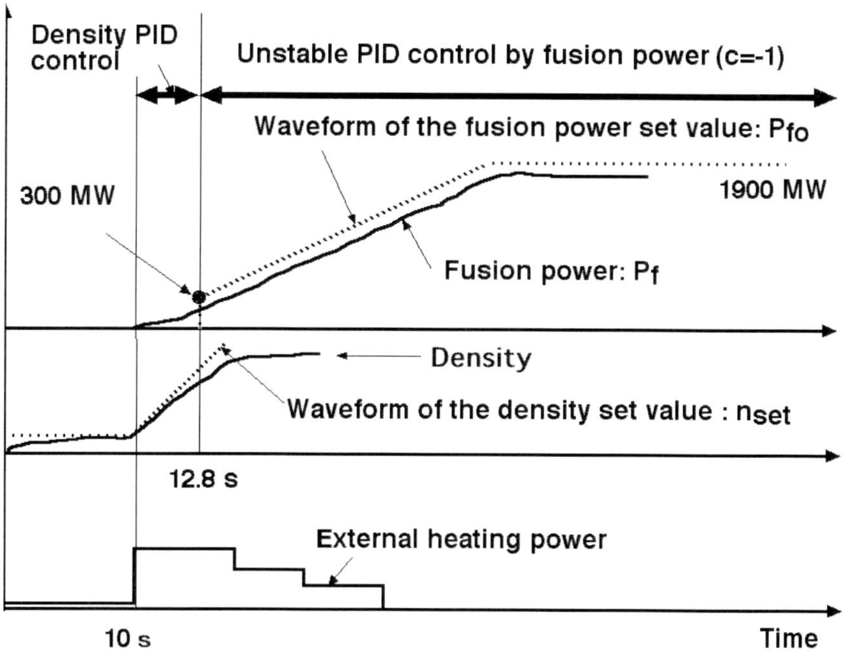

Figure 5. Schematic waveform of the control action based on the density and fusion power.

The temporal evolution of the plasma parameters for continuous fueling is shown in Figure 6. The initial heating power is P_{EXT}=90 MW, the fusion power rise-up time is Δt_{rise}=20 s, and other parameters are γ_{ISS}=1.6, τ_α^*/τ_E=3, τ_p^*/τ_E=3, γ_{SUDO}=5.5, T_d=0 s, and T_{int}=10 s. It is seen that the plasma parameters smoothly changed due to continuous fueling. The peak temperature at the steady state is $T_i(0)$~6.4 keV, the peak density is $n(0)=9.62\times10^{20}$ m^{-3}, the volume averaged beta value is <β>~ 2.5 %, the alpha ash fraction is f_α~5.4 %, the effective charge is Z_{eff}~1.52, the average neutron wall loading is Γ_n~1.5 MW/m^2, the ratio of the Bremsstrahlung loss power P_B to the alpha heating power P_α is P_B/P_α ~ 71 %, reducing the divertor heat load to Γ_{div}~5.4 MW/m^2 for the 0.1 m width of the wetted divertor plate at the right angle to the magnetic field line. This right angle divertor geometry is employed in this study for simplicity unless otherwise noted. Other parameters are also listed in Table 1.

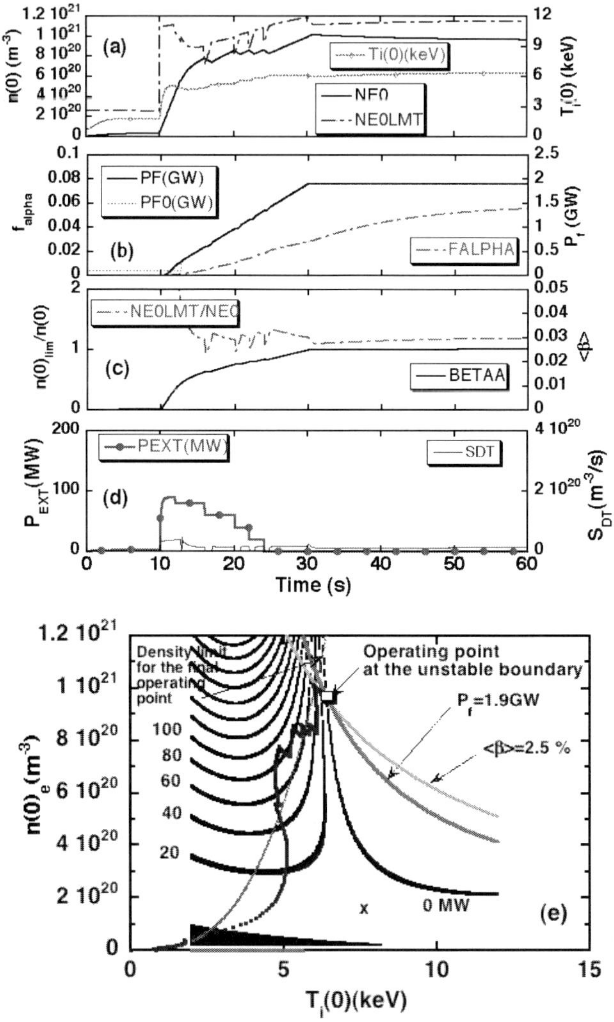

Figure 6. Temporal evolution of the plasma parameters for continuous fueling. (a) Peak temperature T(0), peak density NE0, density limit NE0LMT, (b) alpha ash fraction FALPHA, fusion power PF and its set value PF0, (c) density limit margin, NE0LMT/NE0, beta value BETAA, and (d) D-T continuous fueling rate SDT, and the external heating power PEXT, (e) the operation path on POPCON. Control parameters are T_d=0 s and T_{int}=10 s.

In Figure 6-(e) is shown the POPCON and operation path (n(0)-T(0) trace) accessing the unstable ignited operating point. POPCON is the contour map of the heating power of $P_{HT} = (\bar{P}_L + \bar{P}_B + \bar{P}_S - \bar{P}_\alpha)$, which is obtained in Eq. (2-3) and plotted on the n(0)-T(0) plane. Following POPCON plots for discrete fueling are given for the time averaging alpha ash fraction.

3.2. Small Density Perturbation Corresponding to 12 mm Size Pellet

The density and fusion power control for continuous fueling is conducted as shown in Figure 5 in the following study. Figure 7 shows the temporal evolution of plasma parameters when the pellets with the size of L_p=12 mm (corresponding to S_{DTpell}= 0.99×10^{20} m^{-3} are injected in FFHR2m. For the fusion power rise-up time of $\Delta\tau_{rise}$=20 s and the maximum external heating power of P_{EXT}=90 MW, the time averaged density is initially built up to ~0.6×10^{21} m^{-3} by the density feedback (Nset trace) and then raised up to n(0)~1×10^{21} m^{-3} by the fusion power control, which is switched on at 12.8 s. The external heating power is preprogrammed and set to 0 at 24 s. We see that even by fueling at the discrete time the ignition access is possible. When the density is increased by discrete fueling, the temperature is dropped and the density variation is Δn(0)~ 2.5×10^{19} m^{-3} (Δn(0)/n(0)~2.5 %). Their variations are out of phase due to adiabatic process in a short time. This is because at the equilibrium operating point the input and output power are balanced, and then there is virtually no energy exchange. As it is similar to the adiabatic condition, the plasma energy is almost kept constant between pellet injections. Therefore, the beta value changes smoothly and shows the constant value as shown in Figure 7-(c). The variation of the diverter heat flux is $\Delta\Gamma_{div}/\Gamma_{div}$~0.1/5.5=3.6 %, and variation of the first wall heat flux is $\Delta P_{HF}/P_{HF}$~0.01/0.24=4.1 %.

As the fusion power control is started at t= 12.8 s, setting of the density n_{set} at 13 s is important to perform the smooth transition of the fusion power. Careful adjustment has been conducted to have a smooth transition. To study the overall fusion power rise-up phase, smaller pellet size was used temporarily before 12.8 s. But it should be optimized in future study. At 10 s, the large fueling signal appears. This is due to continuous fueling for a very short time.

In Figure 7-(e) is shown the operation path to the unstable ignition point on POPCON. The operation is stabilized by cooling with fueling and by heating with the fueling reduction, which is controlled by the error of the fusion power $e_{DT}(P_f) = -(1- P_f/P_{fo})$. We see that the operating point never exceed 6.3 keV and the operation regime is in the plateau regime with respect to the collision frequency.

Detailed waveform of the fusion power and fueling rate $S_{DT}(t)$ are shown in Figure 8 during the fusion power rise-up phase of between 15 s and 20 s. As the actual fusion power is lower than the set value ($P_f < P_{fo}$) (Figure 8-(b)), fuel should not be supplied if only the proportional control (ERPF) is employed in the control algorithm (Eq. (2-16)) as shown in Figure 8-(c).

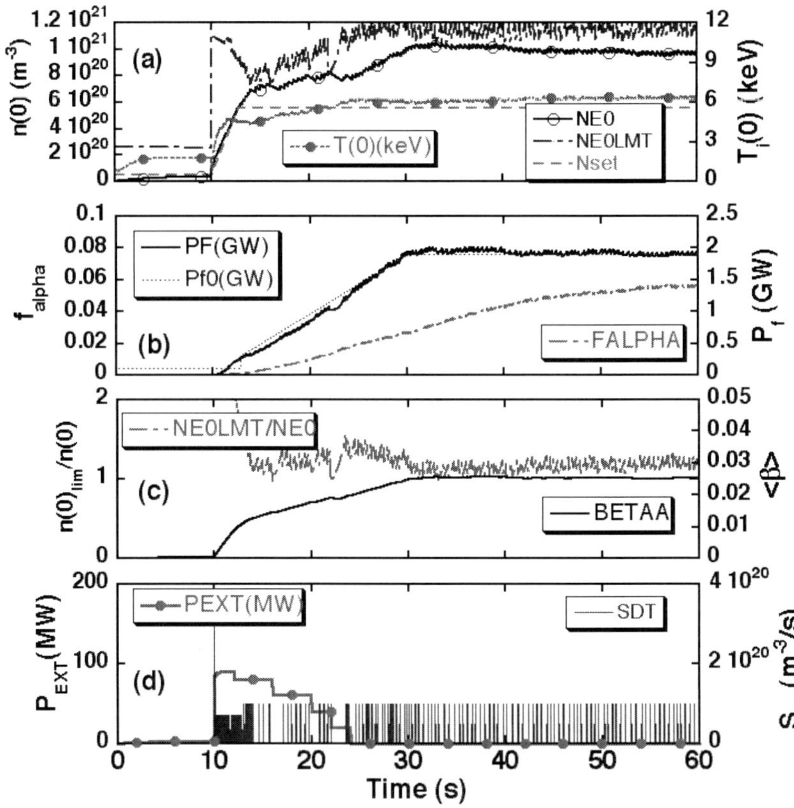

Figure 7. (Continued).

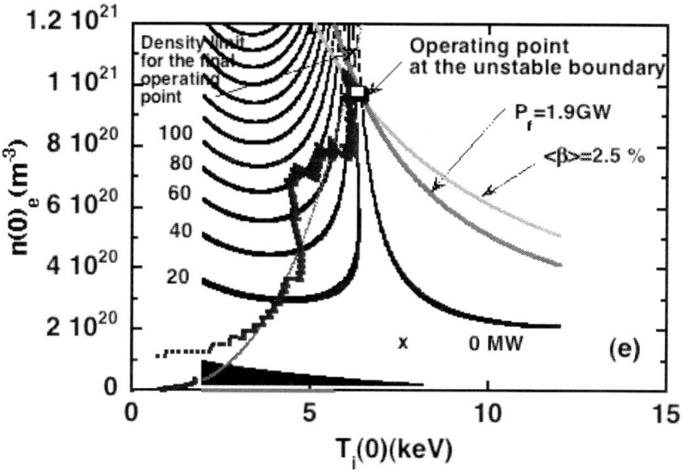

Figure 7. Temporal evolution of the plasma parameters for discrete fueling with pellet size of 12 mm. (a) -(e) are the same as in Figure 6. (T_d=0.26 s and T_{int}=8 s).

Detailed waveform of the fusion power and fueling rate $S_{DT}(t)$ are shown in Figure 8 during the fusion power rise-up phase of between 15 s and 20 s. As the actual fusion power is lower than the set value ($P_f < P_{fo}$) (Figure 8-(b)), fuel should not be supplied if only the proportional control (ERPF) is employed in the control algorithm (Eq. (2-16)) as shown in Figure 8-(c). Therefore, the density should not increase in the phase of $P_f < P_{fo}$. However, ERROR(P_f) in Figure 8-(c) containing the derivative term makes the fueling possible due to the positive derivative of the fusion power with respect to the time, which is clearly seen in the oscillating trace of the fusion power (PF) in Figure 8-(b). As positive values of PID signal of ERROR(P_f) are providing the discrete fueling timing, setting of the derivative time T_d is important in the discrete fueling, which is different from the continuous fueling operation.

The total fuelled particle number for continuous fueling and discrete fueling should be the same in the steady state. For example, averagely 22 pellet trains with $S_{DTpell}=0.9927\text{x}10^{20}$ m^{-3} are injected in the time interval of 3x0.02 s during 10 s in the steady state between 50 s and 60 s as shown in Figure 7-(d). Therefore total fuelled particles are $S_{DTpell}=22\text{x}0.9927\text{x}10^{20}$ m^{-3}x(3x0.02) = 1.31 x10^{20} m^{-3} in 10 s of the steady state phase, yielding S_{DT}=0.13x x10^{20} m^{-3}/s. This is almost equal to $S_{DT}=0.1342\text{x}10^{20}$ m^{-3}/s for continuous fueling as seen in Figure 6-(d). From Eq. (2-1) the peak density can be roughly estimated as $n(0) \sim (1+\alpha_n) S_{DT} \tau_p^* \sim 5\text{x}0.1342\text{x}\ 10^{20}\text{x}\ (4.7\text{x}3) =$

$9.4x10^{20}$ for the confinement time τ_E=4.7 s listed in Table 1 and τ_p*/τ_E =3, which is comparable to the obtained peak density.

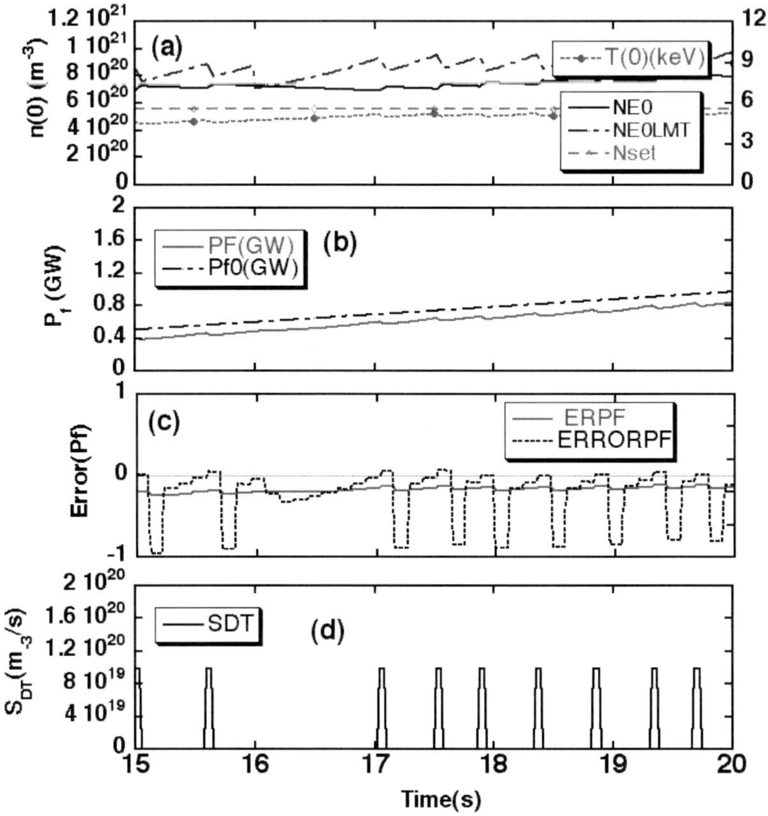

Figure 8. Detailed waveforms of Figure 7. (a) Peak temperature, peak density, and density limit, (b) fusion power and its set value, (c) the error of the fusion power composed of only the proportional term (ERPF), and the PID term (ERRORPF), (d) D-T discrete fueling rate SDT.

3.3. Medium Density Perturbations Corresponding to 14 mm Size Pellet

For the larger pellet size of 14 mm (corresponding to S_{DTpell}= $1.57x10^{20}$ m^{-3}), the fusion power smoothly rises up as shown in Figure 9 although actual fusion power is slightly smaller than the set value for the case of T_d =0.26 s and T_{int}=8 s. This is due to the preprogramming of the external heating power. Therefore, we mainly use 14 mm size pellet injection in FFHR2m study. As shown in Figure 9-(f), the density variation is $\Delta n(0)$~$5x10^{19}m^{-3}$ ($\Delta n(0)/n(0)$~5 %) in the steady state phase. The variation of the diverter heat flux is $\Delta\Gamma_{div}/\Gamma_{div}$~0.3/5.5= 5.4%, and variation of the first wall heat flux is $\Delta P_{HF}/P_{HF}$~0.02/0.24=8.3% as seen in Figure 9-(g).

In Figure 9-(h) POPCON is plotted. The operation path is similar to 12 mm pellet case (Figure 7-(e)), but the operating traces is showing more zigzag due to the density and temperature variations by larger pellet injections.

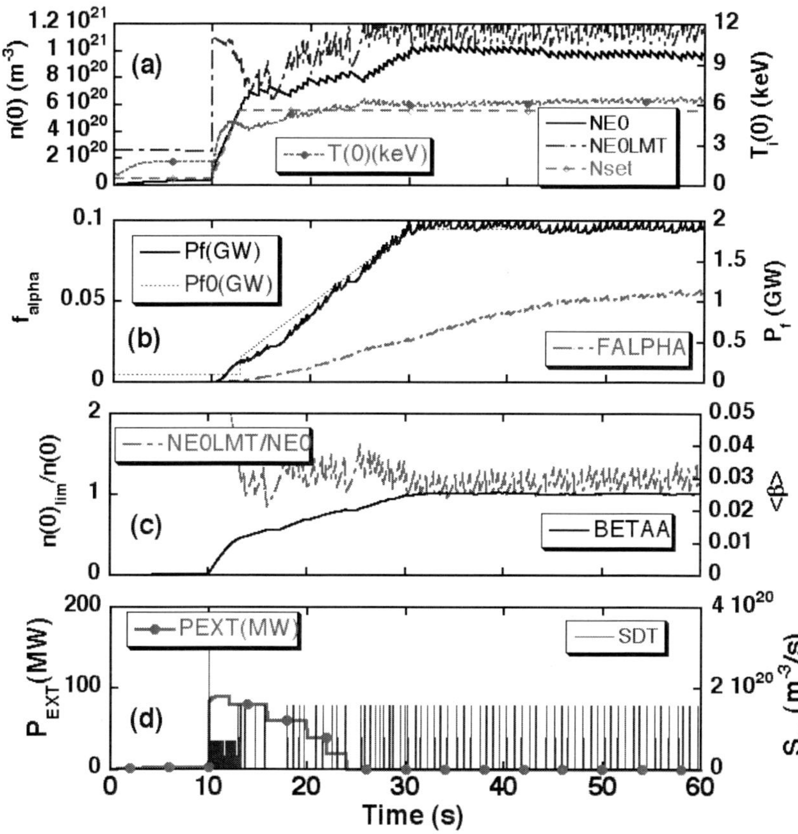

Figure 9. (Continued).

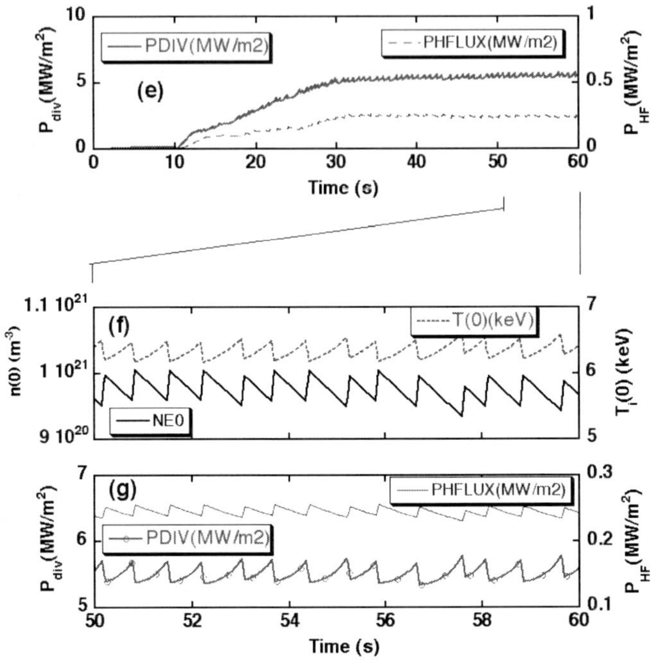

Figure 9. Temporal evolution of the plasma parameters for discrete fueling with the pellet size of 14 mm. (a)-(d) are the same as in Figure 6, (e) the divertor heat flux PDIV and the first wall heat flux PHFLUX, (f) density and temperature variations between 50 s and 60 s, and (g) the divertor heat flux PDIV and the first wall heat flux PHFLUX variations. (T_d=0.26 s and T_{int}=8 s).

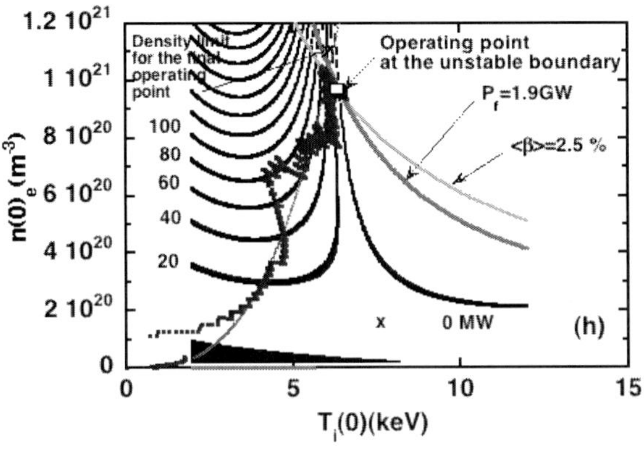

Figure 9-(h). The operation path on POPCON.

This non-model based treatment does not contain any gain in fueling control due to the digitized nature of the discrete fueling during the ignited operation. However, choice of the derivative and integral times is important to achieve the unstable operation with discrete fueling. As detailed operational regime with the respect to T_d and T_{int} parameters was surveyed with the feedback control of the heating power based on the fusion power [23], here we show only a result in the case of no feedback control of the heating power in Figure 10. The solid square shows the parameter of T_{int}=8 s and T_d=0.26 s mainly used in this paper. For the integral time of T_{int}=8 s, the sound fusion power rise-up is obtained when the derivative time is between T_d=0.06 s and 0.5 s. As the calculation time step is Δt=0.02 s, the derivative time T_d should avoid this timing area for activating the derivative term. Therefore, T_d=0.001 s smaller than Δt=0.02 s can be used. For the larger derivative time than T_d =0.6 s, the fusion power does not increase as set, and then terminates at ~25 s. On the other hand, we have found that the integral time between T_{int}=0.4 s and 45 s provides the sound fusion power rise-up for T_d=0.26 s. For the longer integral time than T_{int}= 50 s the fusion power can be increased but is delayed, and finally quenched. We thus see that the control parameter sets of T_{int} and T_d for the thermally unstable operation are quite wide contrary to our expectation. Parameters for the best performance should be in this regime, but not optimized yet.

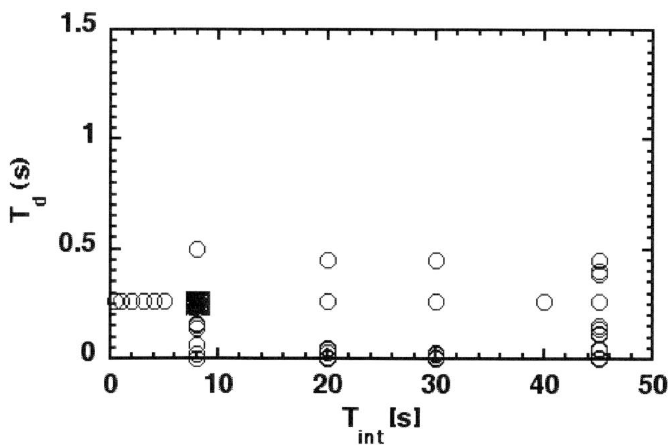

Figure 10. Ignited operation regime for the integration T_{int} and the derivative times T_d for discrete fueling with the pellet size of 14 mm. Open circles indicate the controllable ignited operation and solid square typical value in this study.

3.4. Large Density Perturbations Corresponding to 16 mm Size Pellet

For the larger pellet size of 16 mm (corresponding to S_{DTpell}=2.35x10^{20} m^{-3}), temporal evolutions of the plasma parameters and POPCON are shown in Figure 11 and Figure 11-(g), respectively. Although the fusion power evolution is not smooth and bumpy due to irregular pace of the pellet injection, the steady state is reached even with the large parameter variations. The density variation in the steady state is as large as Δn(0)~ 8x10^{19} m^{-3} (Δn(0)/n(0)~8 %) and the temperature variation is ΔT(0)~ 0.8 keV, but ignition is still maintained. Although the variation of the fusion power is also increased to $\Delta P_f/P_f$ ~6.8 %, we expect that the coolant temperature from the blanket might be smeared out. The divertor heat flux is around 5.4 MW/m^2, and its variation is $\Delta\Gamma_{div}/\Gamma_{div}$~0.5/5.5=9 %. As the divertor heat flux is in the lower level, it may not so problematic. However, thermal fatigue in the divertor plate should be carefully estimated. The first wall heat flux is larger than that in the stable operating point, but it is still in the low level as P_{HF}~0.24 MW/m^2, and the variation of the heat flux to the first wall is $\Delta P_{HF}/P_{HF}$~0.025/0.24=10 %. Thermal fatigue in the first wall should be carefully estimated. We note that as seen in Figure 11-(f) the beta value does not change too much due to adiabatic nature while the fusion power varies sharply. As the temperature and the density change out of phase as seen in Figure 11-(e), the beta value proportional to n(0)T(0) does not change. On the other hand, the fusion power changes so much because of $P_f \sim n^2(0)T^3(0)$.

In POPCON, the operating point oscillates almost in parallel with the constant beta line during the ignition access and near the operating point. During the ignition access, the input power is slightly larger but the dW/dt term is finite and small. Therefore, the operating point is not far from the equilibrium point, and the operating point behaves as adiabatic. Thus, the plasma energy is almost kept constant between the discrete fueling, leading to the movement on the constant beta line. This is also seen on detailed POPCON in Figure 11-(i) where the steady state operating point moves on the constant beta line of $\langle\beta\rangle$=2.57 %. Therefore, the fusion power drops quickly when discrete fueling pulses are provided as shown in Figure11-(g) and (i). As the operating point moves quickly during discrete fueling, it does not make a circle. We note that the same parameters of T_d=0.26 s and T_{int}=8 s are used.

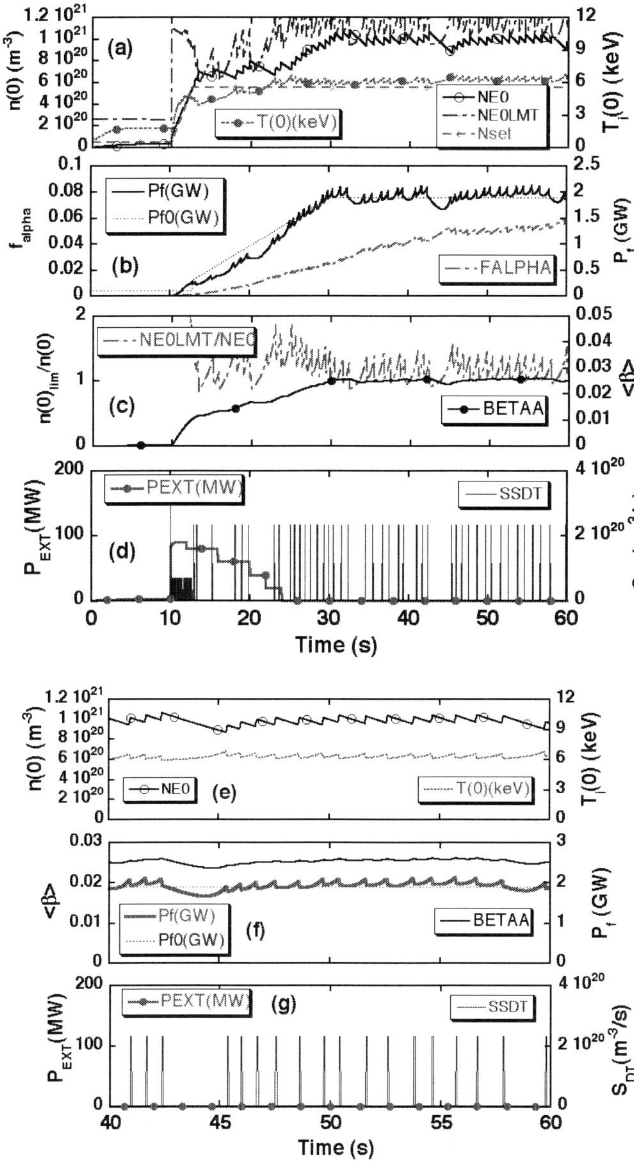

Figure 11. Temporal evolution of the plasma parameters for discrete fueling with the pellet size of 16 mm. (a)-(d) are the same as in Figure 6, (e) density and temperature variation between 40 s and 60 s, (f) the beta value and fusion power variations, and (g) the discrete fueling pulses. (T_d=0.26 s and T_{int}=8 s).

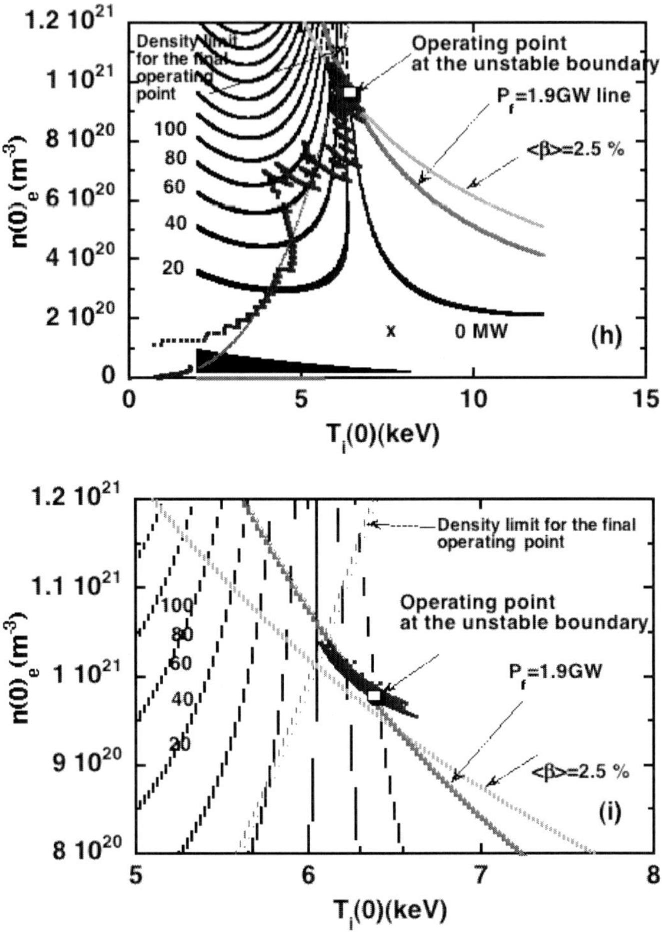

Figure 11-(h) and (h). (h) The operation path on POPCON corresponding to Figure 11-(a), and (i) detailed operating point behavior around the ignition boundary on POPCON between 50 and 55 s.

3.5. Extra-large Density Perturbations Corresponding to 17 mm Size Pellets

For the larger pellet size of 17 mm (corresponding to $S_{DTpell}=2.82 \times 10^{20}$ m^{-3}), the fuelled particle number is increased and the density variation is slightly larger than $\Delta n(0) \sim 8 \times 10^{19} m^{-3}$ ($\Delta n(0)/n(0) \sim 8$ %) in the case of 16 mm pellet size, but ignition is terminated at t=56.5 s as shown in Figure 12.

Control parameters in the fueling are T_d=0.19 s and T_{int}=8 s. After 55 s, both the density and the temperature are decreased at the same time, and their variations are in phase as seen in Figure 12-(f), and then ignition is terminated. As ignition termination cannot be avoided even by the other choice of parameters of T_d and T_{int}, 17 mm pellet size cannot be used for unstable operation in this reactor as long as three-pellet train is employed.

For the smaller pellet size of 10 mm (corresponding to S_{DTpell}=0.5745x10^{20} m^{-3}), it is found that the slow fusion power oscillations take place, and cannot have the constant fusion power for any control parameters of T_d and T_{int}. As further smaller pellet size cannot supply enough fueling, it is impossible to reach the high-density operation regime. Therefore, it is found from above calculations that the pellet size should be in the range of 12 mm~16 mm in FFHR2m.

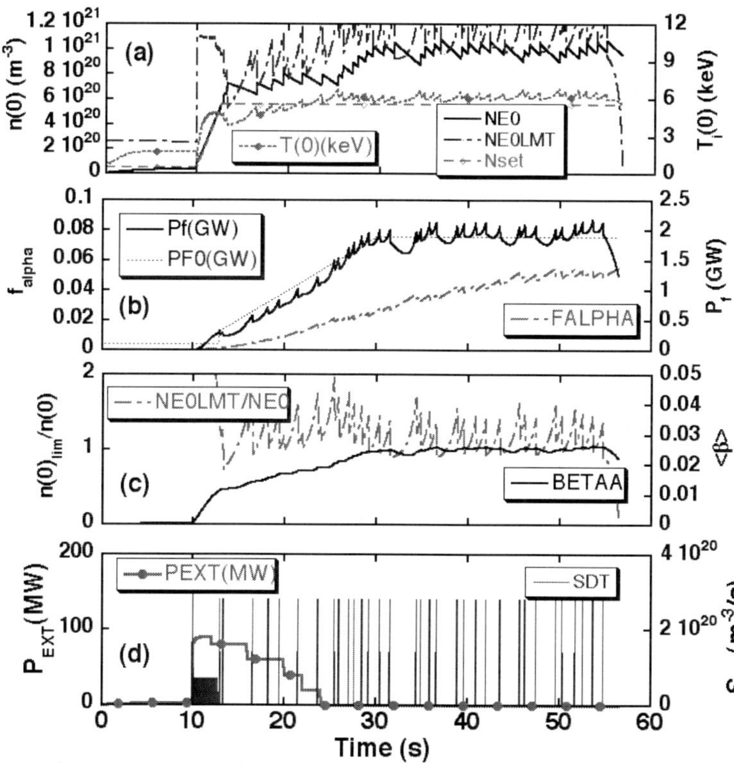

Figure 12. (Continued).

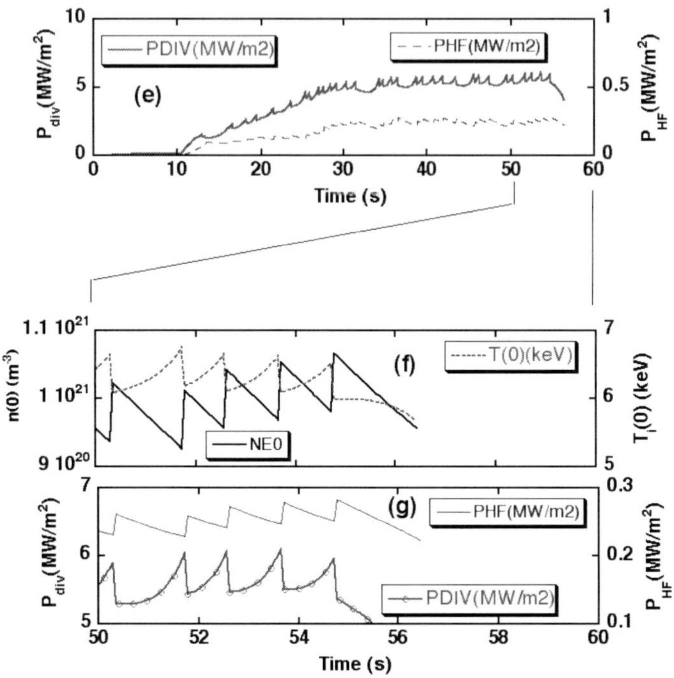

Figure 12. Temporal evolution of the plasma parameters for discrete fueling with the pellet size of 17 mm. (a) -(g) are the same as in Figure 9. Ignition is terminated at 56.5 s. The density and temperature variation are in phase after the last fueling pulse. (Td=0.19 s and Tint=8.1 s).

4. HEATING POWER SPECIFICATION AND ITS FEEDBACK CONTROL IN FFHR-2M2

So far, the external heating method was not specified in high-density ignition access study. In this section we employ the high-energy neutral beam heating (NBI) system with 1.5 MeV, which is most plausible heating method for future use at present [32]. As the electron Bernstein heating (EBW) heating systems may work in the high-density regime, we use this only in the steady state high-density regime. Therefore, in this section we consider the possibility of ignition access by NBI in FFHR2m2 (R=15.7m, $\bar{a}$ =2.5 m, Table 1). During the ignition access, the density is controlled in the range of $n(0)<4x10^{20}$ m^{-3} to ensure the NBI penetration, and then ignition is accessed by alpha heating above this density.

4.1. NBI Penetration Depth

Figure 13 shows the NBI penetration e-folding length for E_{NBI} =1 and 2 MeV calculated by

$$\lambda(n,T)[m] = \left[\int_{-1}^{0} n(x)\sigma_{Yan}(n,T)dx \right]^{-1} \tag{4-1}$$

where σ_{Yan} is the cross section for the multi-step ionization for NBI [33]. Density profile, $n(\rho) = n(0)(1-\rho^2)^{\alpha_n}$, is assumed with ρ being the normalized minor radius. The square root of parabolic α_n=0.5 for the thermally stable operation are assumed, and the peaked one with α_n=3 for the thermally unstable operation. The temperature profile is parabolic with α_T=1 in all the cases.

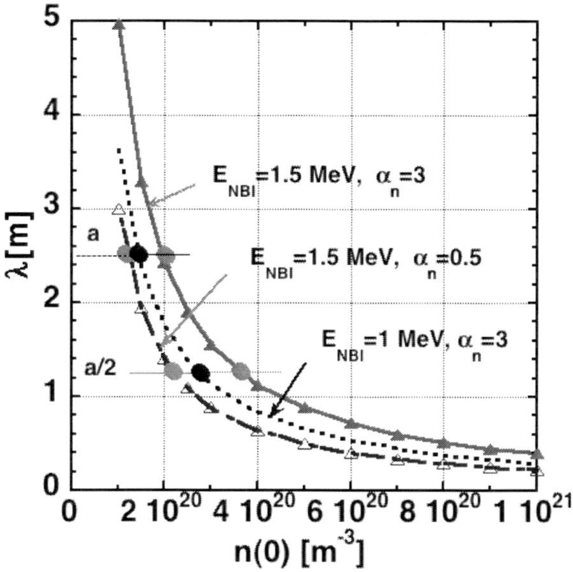

Figure 13. NBI penetration e-folding length with respect to the peak density in FFHR2m2 (R=15.7 m) with the density profile of α_n=3 and α_n=0.5. "a" and "a/2" show a minor radius and a half minor radius. The temperature profile is parabolic (α_T=1) in both cases.

In Figure 13 is shown the NBI penetration length λ in the R=15.7 m reactor with the peaked profile (α_n=3) and broad profile (α_n=0.5) for E_{NBI}=1.0 and 1.5 MeV. λ=2.43 m and 1.13 m are obtained for the density of n(0)~ $2x10^{20}$ m^{-3} and $4x10^{20}$ m^{-3}, respectively, for the peaked profile (α_n=3) and 1.5 MeV. For E_{NBI}=1.0 MeV, they are reduced to λ=1.77 m and 0.834 m at the respective density of n(0)~ $2x10^{20}$ m^{-3} and $4x10^{20}$ m^{-3}. As 1 MeV NBI system is now planned for ITER project [32] and up to 2 MeV is foreseeable, 1.5 MeV NBI has been used in this study. We also employed tangential injection to decrease the initial shine-through power in this study.

To access the ignition regime, the initial density should be low for NBI penetration as seen later. Therefore careful operation is necessary.

4.2. Feedback Control of the External Heating Power

So far the preprogramming control has been used for the external heating power because the suitable control algorithm has not been discovered. However, upon considering the analogy of the density limit for the heating power feedback control to the thermally stable ignition access, the temperature limit concept has been emerged for the suitable heating power control as explained below.

4.2.1. Basic principle of this control algorithm

The thermally stable and unstable operation regime has an inverse relationship with respect to T-n axis in the POPCON. Here, POPCON is the contour map of the heating power of $P_{HT} = \left(P_L + P_B + P_S - P_\alpha\right)$ plotted on the n-T plane. This term is involved in the global power balance equation as follows,

$$\frac{d\overline{W}}{dt} = P_{EXT} - P_{HT} = P_{NET} - P_L \tag{4-2}$$

where $\overline{W}$ is the plasma energy and P_{EXT} is the external heating power, Therefore, the operating point proceeds to the final point on the ignition boundary in POPCON depending on the heating power P_{EXT}.

While the operating point rotates to clockwise direction in the thermally stable regime, it rotates to anti-clockwise direction in the thermally unstable regime as shown in Figure 14. In the thermally stable regime, the density limit nlim restricts the operating point to the low-density regime near horizontal temperature-axis. Analogy of the density limit in the thermally stable regime gives us a hint of this control algorithm. The temperature limit Tlim, if it exists, may restrict the operating point to the low-temperature and high-density regime near the vertical density-axis. Actually in this work we demonstrate that the operating point can be guided to the high-density and the low-temperature regime by feedback control of the hearing power based on the assumed temperature limit.

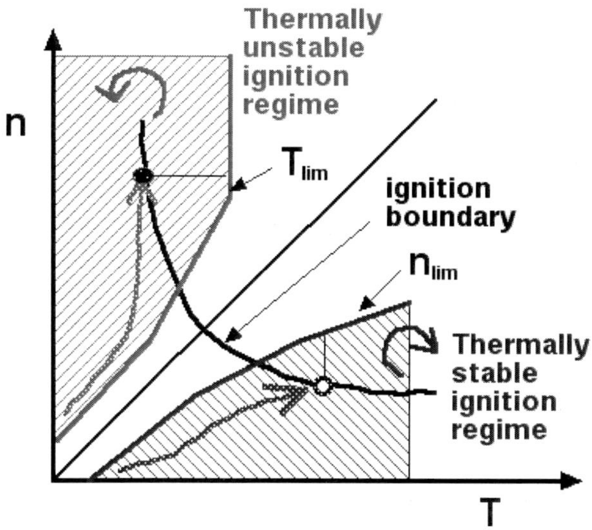

Figure 14. Schematic behaviors of the operating point near the ignition boundary and the density and temperature limit on POPCON. The contour map expressing the heating power P_H is omitted for simplicity.

In the thermally stable low-density and high temperature regime, the Sudo density limit [25] can be used for feedback control of the heating power. It is given below as

$$\gamma_{DLM} n(0) \leq n(0)_{\lim}[m^{-3}] = \gamma_{SUDO} \frac{0.25 \times 10^{20}}{\gamma_{pr}} \sqrt{\frac{\{P_{NET}[W] \times 10^{-6}\} B_o[T]}{\bar{a}^2 R[m]}} \quad (4\text{-}3)$$

where P_{NET} is the net heating power given by $P_{NET} = P_{EXT} + (P_\alpha - P_B - P_S)$ in Eq. (4−2), R is the major radius, $\bar{a}$ is the effective minor radius, B_o is the magnetic field strength, γ_{SUDO}=1.5 is the density limit factor, and γ_{pr} is the density profile factor. The external heating power for feedback control is thus obtained from this inequality as [27]

$$P_{EXT}[W] \geq \left\{ \gamma_{pr} \frac{\left[\gamma_{DLM} n(0)[m^{-3}]\right]}{\gamma_{SUDO} 0.25 \times 10^{20}} \right\}^2 \frac{\bar{a}^2 R[m]}{B_o[T]} \times 10^6 - \left(P_\alpha - P_B - P_S\right) \tag{4-4}$$

Actually as shown in Figure 15, the feedback controlled heating power works well during the ignition access to the thermally stable regime, and is robust to the various disturbances. The operating density is always smaller than the density limit. When the confinement factor is reduced, the heating power becomes finite, yielding sub-ignited operation.

As an analogy of the density limit concept, we postulate the temperature limit restricting the operation regime to the low temperature regime as,

$$\gamma_{TLM} T(0) \leq T(0)_{\lim} = T_c \left[\frac{\left\{P_{NET}[W] \times 10^{-6}\right\} B_o[T]}{\bar{a}^2 R[m]} \right]^{\alpha_{TC}} \tag{4-5}$$

where α_{TC} and T_c was determined by simulations, and γ_{TLM} is set to 1.2 to have a higher temperature limit $T(0)_{lim}$ than the operating temperature T(0). Rewriting this inequality provides the external heating power for a feedback purpose as,

$$P_{EXT}(T_{LIM}) \geq \left\{ \frac{\gamma_{TLM} T(0)}{T_c} \right\}^{\frac{1}{\alpha_{TC}}} \frac{\bar{a}^2 R[m]}{B_o[T]} \times 10^6 - \left(P_\alpha - P_B - P_S\right) \tag{4-6}$$

To execute this control action, it is necessary to measure the temperature T(0) during the ignition access in addition to the power balance terms and T_c=7.4 keV and α_{TC}=0.2 are used unless otherwise noted.

We have used above fueling rates (Eq.(2-14 or 2-17)) and the heating power (Eq.(4-6)) in zero-dimensional particle and power balance equations.

4.2.2. Feedback control of the NBI heating power to the high-density ignition regime with continuous fueling

We have calculated the ignition access to the thermally unstable regime in FFHR2m2 (R= 15.7 m, a=2.5 m) using NBI tangential injection with 1.5 MeV.

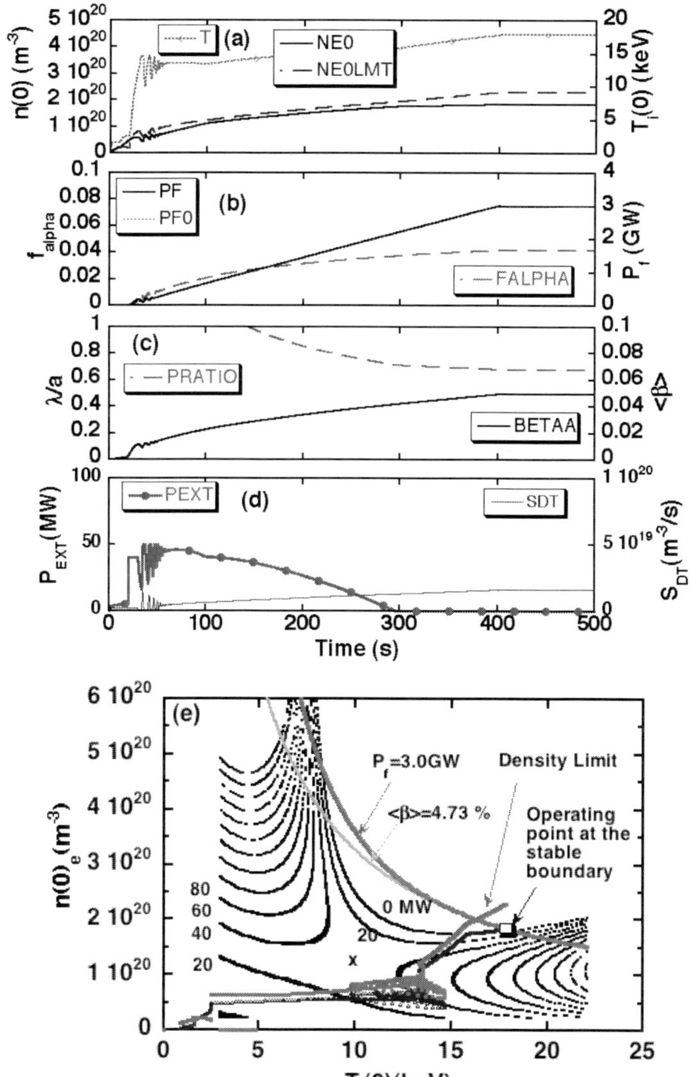

Figure 15. Ignition access to the stable operating point using the feedback control of the external heating power for continuous fueling in FFHR2m2. (a)-(e) are the same as in Figure 6 except for (c) NBI penetration ratio PRATIO. The external heating power is limited to 50 MW.

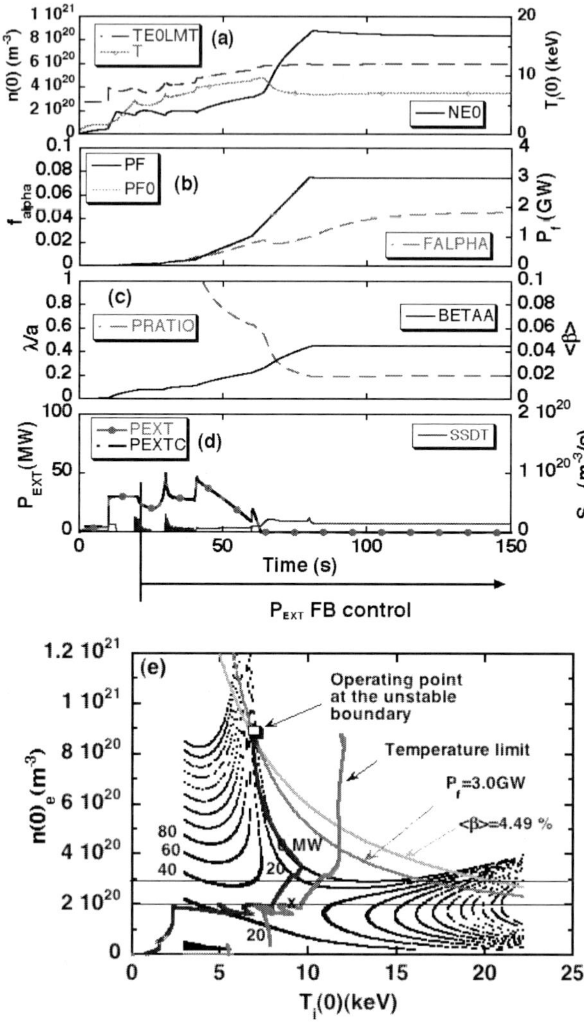

Figure 16. Ignition access to the unstable operating point using the feedback control of the external heating power for continuous fueling in FFHR2m2. (a) Peak temperature T, temperature limit $T(0)_{lim}$ given by Eq.(4-5) TEOLMT and peak density NEO, (b) alpha ash fraction FALPHA, fusion power PF and its set value PF0, (c) NBI penetration ratio PRATIO, beta value BETAA, and (d) D-T continuous fueling rate SDT, and the heating power PEXT, which is limited to 50MW. (e) The operation path on POPCON. Control parameters are T_d=0 s, T_{int}=10 s. We note that "P_{EXT} FB control" below box (d) is describing the feedback control action of the external heating power after 20 s.

The density profile is peaked as α_n=3, temperature profile is parabolic as α_T=1.0 in this calculation. Using this control algorithm, the ignition access can be done instead of the preprogrammed heating power as shown in Figure 16. For T_c= 7.4 keV and α_{TC}=0.2, after 20 s feedback is switched on, and the heating power is automatically applied to keep the temperature T(0) below $T(0)_{lim}$, satisfying inequality Eq.(4-5) (Figure 16-(a)). During NBI heating phase, the NBI penetration ratio to the minor radius is over 0.6 (Figure16-(c)), ensuring the NBI heating capability. As the operating point passes near the saddle point as shown in POPCON in Figure16-(e), this temperature limit setting is close to the optimum value. After reaching ignition regime as seen on this POPCON, the operating point proceeds to the steady state point with the high-density and low temperature regime of n(0)=8.40x10^{20} m^{-3}, T(0)=7.1 keV and <β>=4.5 %. The divertor heat flux is P_{div}=12.1 MW/m^2 for the normal incident angle to the divertor target plate width of 0.1 m.

When T_c is reduced, the actual temperature is forced to be lower. Therefore, the operating point goes along the higher contour line on POPCON, leading to the larger heating power and worse NBI penetration due to higher density.

4.2.3. Feedback control of the NBI heating power to the high-density ignition regime with discrete fueling

Instead of continuous fueling, we study the ignition access by the discrete fueling as shown in Figure 17. The time step of the calculation was the same as Δt=20 ms (T_{rep}=120 ms). Continuous fueling was conducted until 26 s, and then was switched to discrete fueling corresponding to 12mm pellet from 26 to 65 s during the relatively lower density phase of 2x10^{20} m^{-3} for NBI penetration. For the subsequent higher density phase, discrete fueling corresponding to 20mm pellet was applied after 65 s. The feedback controlled external heating power was switched off automatically at ~65 s during the small discrete fueling phase. Although the external heating power waveform is slightly different due to the discrete fueling, waveforms of the temporal evolution are totally similar to continuous fueling case in Figure 16. It is now confirmed that the feedback control algorithm of the external heating power works well in the discrete fueling case. We also notice that at least two sizes of the pellet should be prepared depending on the operation density to have smooth ignition access.

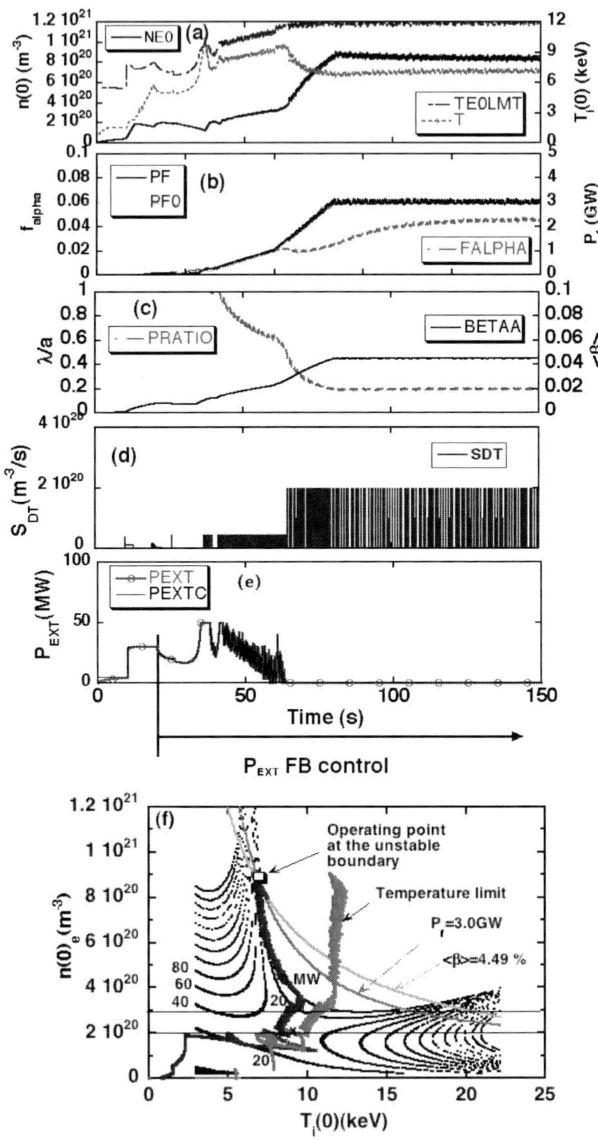

Figure 17. Ignition access to the unstable operating point using the feedback control of the external heating power for discrete fueling with the time step of Δt=20 ms. (a)-(c) are the same as in Figure16 except for (d) D-T discrete fueling rate SDT corresponding to 12 and 20 mm pellets. Second discrete fueling using 20 mm pellet was switched on at 65 s. (e) The external heating power PEXT, which is limited to 50 MW. (f) The operation path on POPCON.

We note that we use the control parameters of T_d=0 s and T_{int}=10, but they are not optimized yet. We also note that if the calculation time step is taken as longer as Δt=50 ms, the operating point sometimes moves out to the high temperature branch depending on the application time and pellet size, leading to the dangerous situation with the large fusion power. This takes place only in the calculation with the longer time step.

5. Control Robustness Against the Disturbances in FFHR2m2

Plasma parameters are inevitably oscillating near the steady state point due to discrete fueling. In addition when the plasma parameter changes largely, this control algorithm should work properly to get the operating point back to the steady state oscillating value. We check here whether this control algorithm works properly against the disturbances for continuous and discrete fueling cases.

5.1. Disturbance by the Change in the Confinement Factor

We already confirmed that when the confinement factor decreases during the heating phase, the external heating power is automatically increased, and vise versa [22]. When the confinement factor is improved suddenly in the steady state phase, fueling increases the density by feedback control to keep the fusion power at the constant value. The heating power is not applied in this case. Therefore we study here the more complicated case when the confinement factor is drooped suddenly in the steady state phase.

In the following calculation, to enhance the sensitivity of the heating power algorithm (Eq. (4-6), the temperature limit factor T_c is set to the lower value of 5.5 keV after the fusion power reaches the steady state. Therefore, the heating power is applied when the temperature is slightly increased. For continuous fueling case, when the confinement factor γ_{ISS} is dropped from 1.43 to 1.27 at 110 s (Figure 18-(c) upper column), the density and the fusion power are decreased quickly, on the other hand the temperature is increased to 11.5 keV as shown in Figure 18-(a) (upper column). This is because the operating point moves to the lower density and higher temperature side on POPCON due to shrinking of the ignition regime. Therefore, the external heating power is applied due to increase in the temperature T(0) in Eq. (4-6). This can be well understood when the temperature ratio $T(0)/T(0)_{lim}$ calculated by two signals in Figure18-(a) is plotted as shown in Figure 18-(d). When the temperature ratio $T(0)/T(0)_{lim}$ is increased sharply after the confinement factor drop, the heating power is applied. This heating power returns the fusion power to the steady state value around 3 GW quickly. After confinement factor drop the density keeps the lower value of $7.1x10^{20}$ m^{-3}, the bremsstrahlung power is lower, and the temperature is higher during lower confinement factor phase until 130 s.

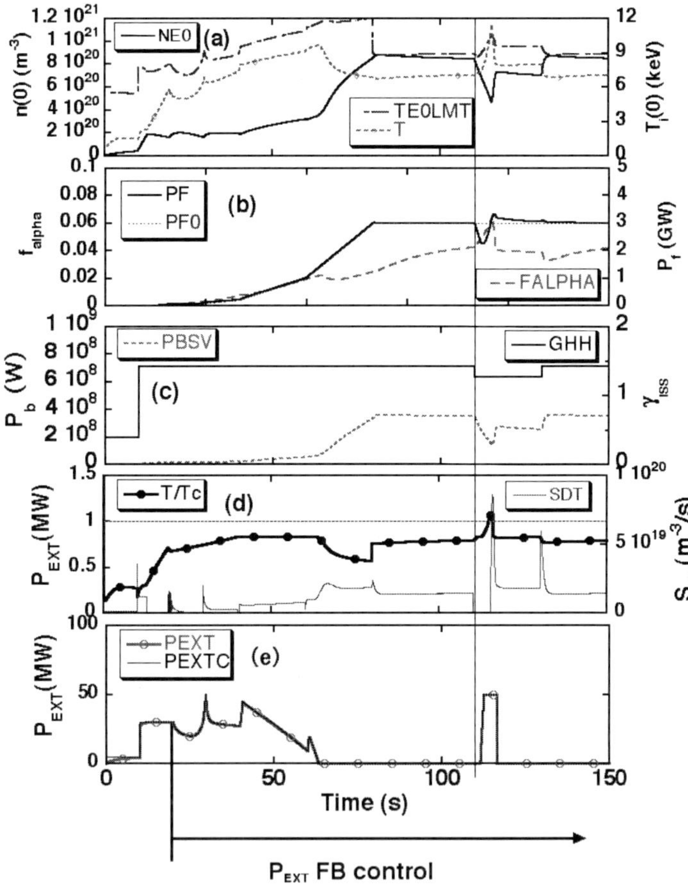

Figure 18. (Continued).

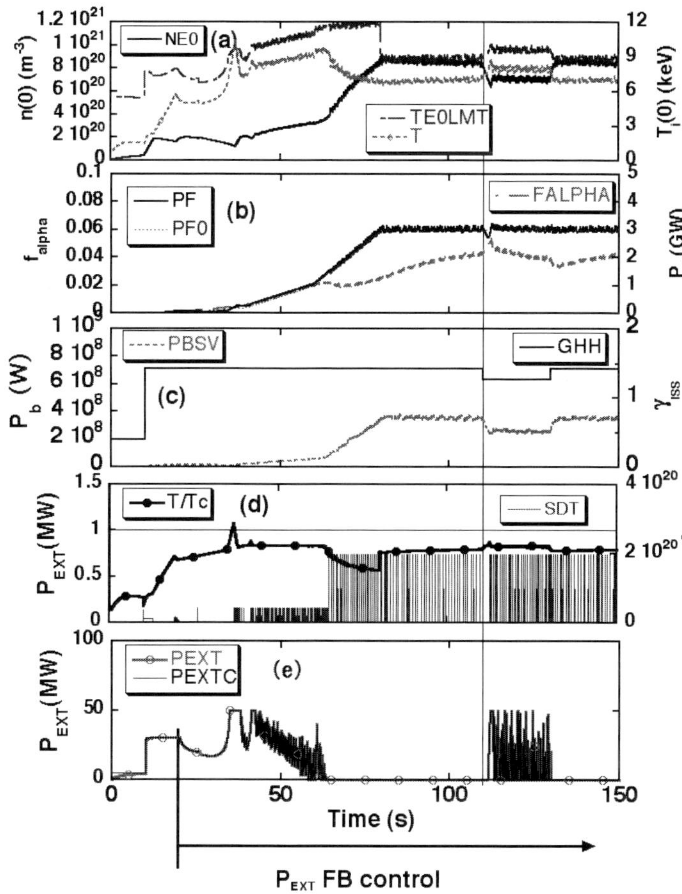

Figure 18. Control effect on the discharge for continuous (upper column) and discrete (lower column) fueling when the confinement factor γ_{ISS} is dropped from 1.43 to 1.27 at 110 s. (a)-(e) are the same as in Figure17 except for (c) the confinement factor GHH, and the bremsstrahlung power PBSV, and (d) the temperature ratio $T(0)/T(0)_{lim}$ and fueling rate SDT. The heating power is limited to 50 MW and T_C =5.5 keV is used after 80s.

When the confinement factor returns to the initial value of 1.43, the density and the temperature return to the previous steady state oscillating values. We also found from this result that the sub-ignited operation is possible using this external heating power (Eq.(4-6) based on the temperature limit.

For discrete fueling case, when the confinement factor is dropped to 1.27, the heating power is applied as well as the continuous fueling case as shown in Figure 18 (lower column), and the discharge can be maintained. The smaller increment of the temperature ratio $T(0)/T(0)_{lim}$ after the confinement factor drop provides the smaller heating power as shown in Figure 18-(e) (lower column). It is thus found that the heating power control algorithm given by Eq. (4-6) works properly against the confinement disturbances to some extent in the steady state by lowering the temperature limit factor T_c. Therefore, when the confinement factor is decreased, the ignited operation could be moved to the sub-ignited operation. However, the time delay is observed in the heating power when the confinement factor changes in both cases.

This time delay of the heating power is not good to cope with disturbances in the steady state situation. Therefore, to improve this situation we introduce the additional feedback control of the external heating power determined by the fusion power error as

$$P_{EXT}(P_f) = P_{EXTO}\left\{ e_{EXT}(P_f) + \frac{1}{T_{P\,\mathrm{int}}}\int_0^t e_{EXT}(P_f)dt \right\} \tag{5-1}$$

where $e_{EXT}(P_f) = (1 - P_f/P_{fo})$, P_{EXT0}=1000 MW and the set value of this fusion power is chosen to be smaller as P_{fo} =2.9 GW than 3.0 GW to prevent fusion power oscillations. We note that the formula was already used in the ITER ignition study [34]. Eq. (5-1) is the proportional-integral (PI) control algorithm to prevent the offset value. T_{Pint}=10 s is used here as in the PID fueling control. In the calculation we compare the heating powers from Eq. (4-6) and Eq. (5-1),

we apply the larger heating power. Therefore the actual heating power is the combined heating power of

$$P_{EXT} = Max\left[P_{EXT}(P_f),\ P_{EXT}(T_{LIM}) \right] \tag{5-2}$$

Using this algorithm, worse condition can be controlled.

We consider the discharge waveform where the confinement factor is dropped from γ_{ISS}=1.43 to the further down value of 1.16 for continuous fueling and to 1.18 for discrete fueling. Using this combined heating power (Eq.5-2), the discharge can be maintained for both the cases as shown in Figure 19 (upper column) for continuous fueling and Figure 19 (lower column) for discrete fueling. Although condition is worse than the previous case (Figure 18), it is working properly by the heating power based on the fusion power error. Here the heating power was limited at 100 MW in the steady state phase for both cases. It is seen that continuous fueling can handle the operation slightly better against confinement disturbance due to quick response of the fueling. On the other hand, the discrete fueling case can still handle the confinement disturbance down to 1.18. Overall, both the fueling methods are not much different to the confinement factor disturbance.

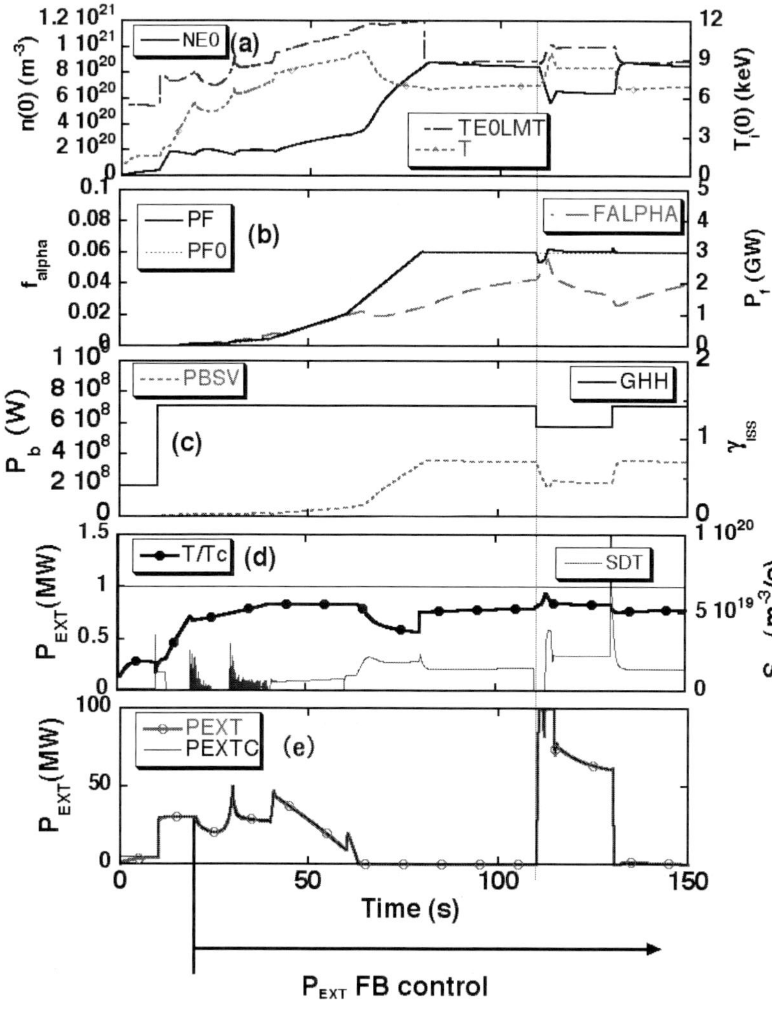

Figure 19. (Continued).

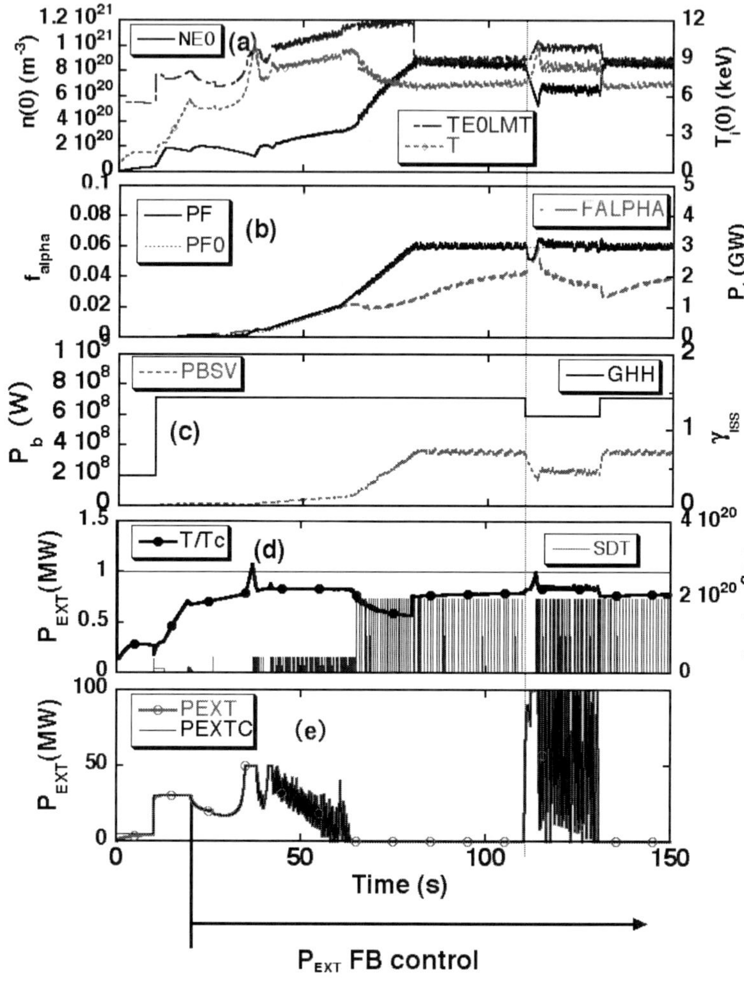

Figure 19. Control effect on the discharge for continuous (upper column) and discrete (lower column) fueling when the confinement factor is dropped from 1.43 to 1.16 and 1.18 at 110 s, respectively. (a)-(e) are the same as in Figure18. The heating power is limited to 100MW in the steady state phase.

We should note here that the external heating system in the steady state to cope with the disturbances should work in the high-density regime. Therefore NBI used in the fusion power rise-up phase cannot be used due to poor penetration. The EBW system is a candidate for this heating system in the high-density regime.

5.2. Disturbance by Impurity Increment

When the impurity contents f_o is increased suddenly, the plasma response is different from the case of the sudden change in the confinement factor. As the bremsstrahlung loss is increased by impurity injection, the temperature increment is smaller than the case with confinement factor drop. Therefore, we have to check whether this feedback control algorithm works properly or not.

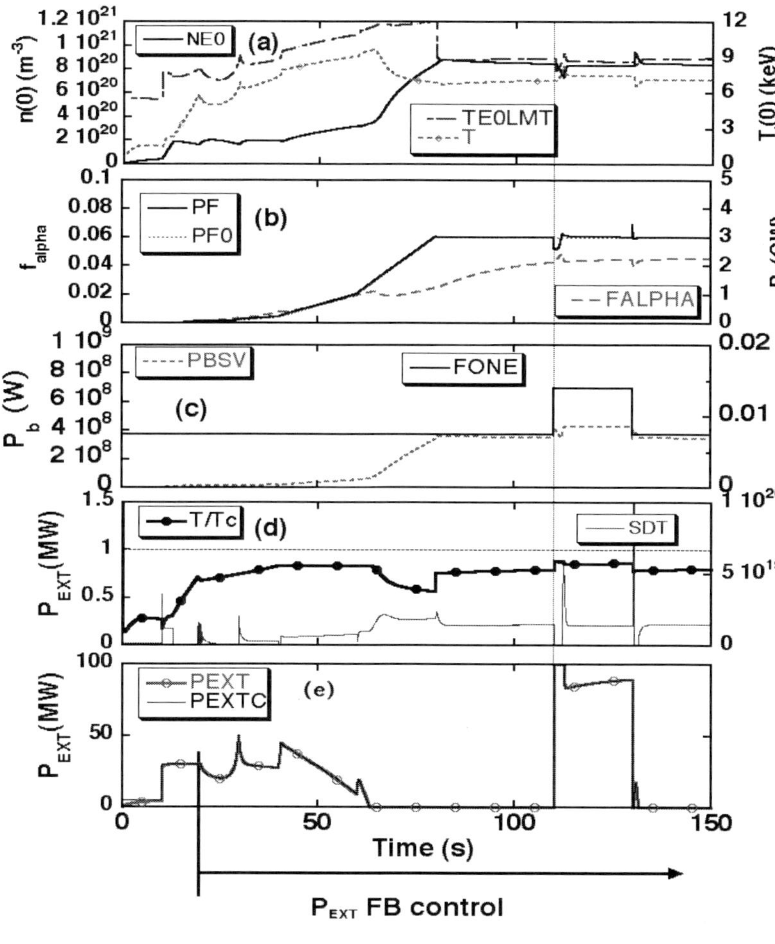

Figure 20. (Continued).

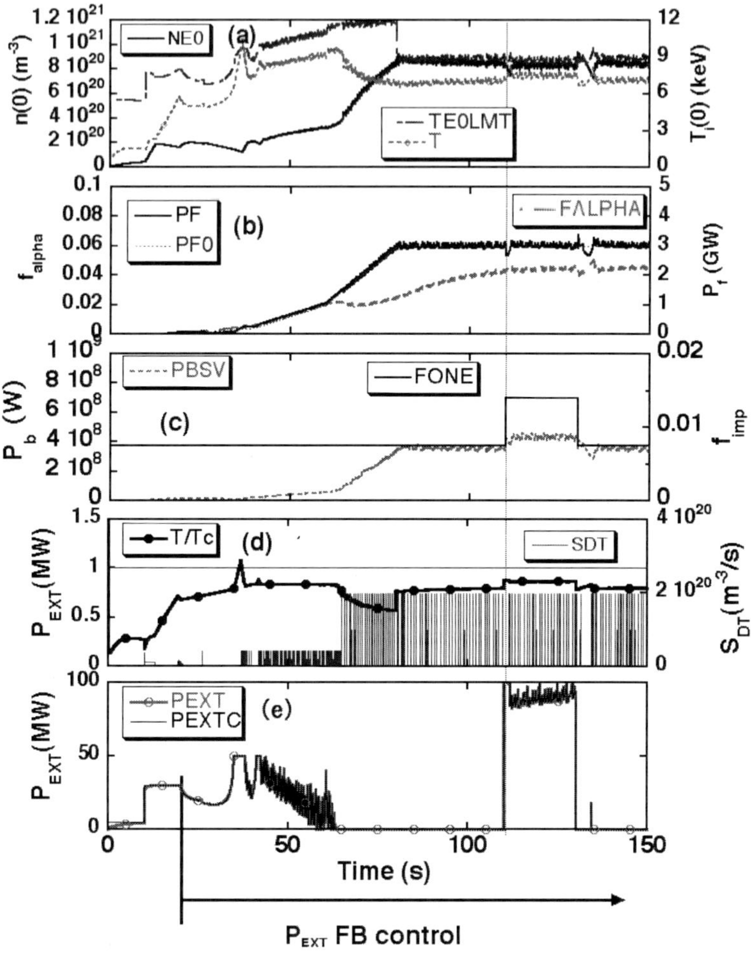

Figure 20. Control effect on the discharge for continuous (upper column) and discrete (lower column) fueling when the impurity fraction is increased from 0.75 % to 1.4 % at 110 s. (a)-(e) are the same as in Figure18 except for (c) the impurity fraction FONE, and the bremsstrahlung power PBSV. The heating power is also limited to 100MW in the steady state phase.

In Figure 20 is shown the response of the discharge against the impurity increment for continuous and discrete fueling. The combined heating power of Eq. (5-2) has been used to keep the fusion power at constant. However when the impurity fraction increases suddenly, the density does not decrease quickly as observed when the confinement factor is dropped suddenly. Therefore, as the delay effect of the heating power is not observed as in Figure 18, then the Eq. (4-6) alone can be used to keep the fusion power at constant. This is seen on the temperature ratio traces $T(0)/T(0)_{lim}$ as shown in Figure 20-(d). As the results with Eq. (5-2) and Eq. (4-6) are the same, the combined heating power (Eq. (5-2)) can be always used in the steady state phase.

6. Discussion and Summary

For simulating the super-dense core discharge in LHD experiments, a box type density and broad temperature profiles are assumed in FFHR2m and peaked density profiles with parabolic shape in FFHR2m2. Before simulation of the discrete fueling such as pellet injection, ignition access by continuous fueling has been established, which provides the basic operation scenarios with respect to the plasma parameters. Continuous fueling can provide the constant values of the plasma parameter in the steady state, but in the discrete fueling such as pellet injection the parameters are oscillating in the steady state. In this 0-dimensional approach we assume that the pellet is completely absorbed in the plasma without the time delay, and the density and temperature profiles are fixed and only their peak values change. However when the pellet is injected as calculated in Appendix-2, the density and temperature profiles change locally. Thus, an interesting problem arises within the 0-dimensional analysis such that how an ignition regime is sensitive to the local profile change. If this problem were answered, it would be helpful for future 1-dimensional analysis.

Of course if the pellet injector has a failure, the thermally unstable operation cannot be stabilized and thermal runaway takes place. Therefore, its backing up system is very important. Such failure mode analysis should be conducted in the future study. However, we believe that the beta collapse would terminate the discharge automatically as the thermal runaway takes place. The divertor and first wall heat flux also vary due to discrete fueling. Therefore, estimation of the thermal fatigue of the blanket and divertor plate is necessary, which is beyond this work.

In principle, the control method described in this paper can be extended to the 1-D simulation and real experimental situations, because only the signals of the total fusion power, total power balance terms and the temperature are necessary for control and hence it is inherently robust against the model uncertainties. In this sense without this pilot study on 0-dimensional analysis, it may be difficult to study the thermally unstable ignition regime using 1-

dimensional simulation code. Future 1-dimensioal analyses can be efficiently and correctly undertaken in collaboration with this 0-dimensioal approach.

To ease the pellet injection, the lower temperature (T(0)< 7 keV) is desirable. To achieve this condition, alpha ash confinement time ratio should be smaller than τ_α^*/τ_E=4~5 and the confinement factor over the ISS95 scaling law of γ_{ISS}>1.4 may be necessary. Helium exhaust experiments and confinement improvement are further necessary in SDC experimental regime in LHD. In this work, we have shown that both the continuous and discrete fueling can stabilize the thermally unstable ignited operating point in FFHR2m, despite its oscillations due to the discrete fueling such as pellet size of 12 and 16 mm size, and the density varies with Δn(0)/n(0)=2.5~8 %. For the larger density variation than Δn(0)/n(0)~8 %, corresponding to the pellet size larger than 17.0 mm, ignition cannot be maintained due to excessively large density variations. Furthermore, the feedback controlled NBI is newly employed as a main heating system in FFHR2m2. Here the initial density is kept at the low value less than ~4x10^{20} m^{-3} to overpass the saddle point on POPCON by NBI heating. After passing the saddle point, a high-density ignition is accessed by alpha heating after turning off NBI injection. This feedback control algorithm of the heating power has been invented based on the temperature limit concept, and it can be used for whole discharge duration, and for various disturbances such as impurity increment and the change in the confinement factor. It is also found that two types of pellets of 12 mm and 20 mm are required in FFHR2m2, and this feedback control method for the heating power is compatible with discrete fueling. Thus, we believe to our knowledge that the unresolved issued on the control of thermally unstable operation pointed out at 1969 has been solved.

ACKNOWLEDGMENTS

This work is performed with the support and under the auspices of the NIFS Collaborative Research Program NIFS14KERP004.

APPENDIX-1. PARTICLE AND POWER BALANCE EQUATIONS

D, T, alpha ash particle balance equations and the electron density using the charge neutrality condition, $n_e(0) = n_D(0) + n_T(0) + 2n_\alpha(0) + 8n_O$, are given by, respectively,

$$\frac{dn_D(0)}{dt} = (1+\alpha_n)S_D(t) - (1+\alpha_n)n_D(0)n_T(0)\overline{<\sigma V>_{DT}(x)} - \frac{n_D(0)}{\tau_D^*} \quad \text{(A-1)}$$

$$\frac{dn_T(0)}{dt} = (1+\alpha_n)S_T(t) - (1+\alpha_n)n_D(0)n_T(0)\overline{<\sigma V>_{DT}(x)} - \frac{n_T(0)}{\tau_T^*} \quad \text{(A-2)}$$

$$\frac{dn_\alpha(0)}{dt} = (1+\alpha_n)n_D(0)n_T(0)\overline{<\sigma V>_{DT}(x)} - \frac{n_\alpha(0)}{\tau_\alpha^*} \quad \text{(A-3)}$$

$$n_e(0) = \frac{n_D(0) + n_T(0) + 2n_\alpha(0)}{1-8f_o} \quad \text{(A-4)}$$

with S_D being deuterium fueling rate, S_T being tritium fueling rate, τ_D* being the effective deuterium confinement time, τ_T* being the effective tritium confinement time, $\overline{\langle\sigma V\rangle_{DT}(x)}$ being the volume averaged D-T fusion rate, and f_o=n_o/n_e(0) being oxygen impurity fraction. In this study we assumed the time constant and flat impurity profiles for the simplicity. For more detailed treatment, the impurity particle balance equations should be used with the impurity particle confinement time. The particle balance equation (Eq. (A-4)) for the electron density n_e(0) is thus given by

$$\frac{dn_e(0)}{dt} = \frac{1}{1-8f_O}\left[(1+\alpha_n)S_{DT}(t) - \left\{\frac{f_D + f_T}{\tau_p^*} + \frac{2f_\alpha}{\tau_\alpha^*}\right\}n_e(0)\right] \quad \text{(2-1)}$$

where S_{DT}=S_D+S_T, τ_p*=τ_D*=τ_T*, f_D=n_D(0)/n_e((0), f_T=n_T(0)/n_e(0), and the alpha ash fraction f_α= n_α(0)/n_e(0). The alpha ash fraction is rewritten from Eq. (A-3) using $n_\alpha(0) = f_\alpha n_e(0)$ as

$$\frac{df_\alpha}{dt} = (1+\alpha_n) n_e(0) f_D f_T \overline{\langle \sigma V \rangle_{DT}(x)} - \frac{f_\alpha}{\tau_\alpha^*} - \frac{f_\alpha}{n_e(0)} \left[\frac{dn_e(0)}{dt} \right] \tag{2-2}$$

The 0-dimensional global power balance equation Eq.(2-3) is

$$\frac{d\overline{W}}{dt} = P_{EXT} / V_o - \left(\overline{P}_L + \overline{P}_B - \overline{P}_\alpha \right) \tag{2-3}$$

where $\overline{W}$ is the volume averaged plasma energy given by

$$\begin{aligned} \overline{W} &= 1.5k\overline{\left(n_D + n_T + n_\alpha + n_e \right) T_i}[^\circ K] \\ &= \frac{1.5 \times 1.6 \times 10^{-19} \left(f_D + f_T + 1/\gamma_i + f_\alpha \right) n_e(0) T_i(0)[eV]}{\left(1 + \alpha_n + \alpha_T \right)} \end{aligned} \tag{A-5}$$

Here, k is the Boltzman constant, $\gamma_i = T_i(0)/T_e(0) = 1$, $e = 1.6\text{x}10^{-19}$ [C]. By differentiating the plasma energy $\overline{W}$ in the power balance equation, Eq. (2-3), the ion temperature equations is obtained as

$$\begin{aligned} \frac{dT_i(0)}{dt} = & \frac{1+\alpha_n+\alpha_T}{1.5e\left(f_D + f_T + 1/\gamma_i + f_\alpha \right) n_e(0)} \times \left[\left\{ P_{EXT}/V_o + \overline{P}_\alpha \right\} - \left\{ \overline{P}_L + \overline{P}_b + \overline{P}_s \right\} \right] \\ & - \frac{T_i(0)}{\left(f_D + f_T + 1/\gamma_i + f_\alpha \right)} \times \left[\left\{ 1 - 8 f_O + \frac{1}{\gamma_i} - f_\alpha \right\} \frac{1}{n_e(0)} \left[\frac{dn_e(0)}{dt} \right] - \left[\frac{df_\alpha(0)}{dt} \right] \right] \end{aligned} \tag{2-4}$$

for the temperature profiles $T_i(x)/T_i(0) = T_e(x)/T_e(0) = (1-x^2)^{\alpha_T}$.

Using the ISS95 confinement time scaling with S=0.59

$$\begin{aligned} \tau_E[s] &= \frac{\gamma_{ISS} \times 0.079 \iota_{2/3}^{0.4} \overline{n}_{19}^{0.51} [\times 10^{19} m^{-3}] B_o^{0.83} [T] \overline{a}^{2.21} R^{0.65} [m] / \left(2\pi R \pi \overline{a}^2 \right)^S}{\left\{ \left(P_{NET} - d\overline{W}/dt \right) 2\pi R \pi \overline{a}^2 \times 10^{-6} \right\}^S [MW] \acute{Z}} \\ &= \frac{C_{op}'}{\left(P_L \times 10^{-6} \right)^S} \end{aligned} \tag{A-6}$$

Here, the power balance equation, Eq. (2-3), is used in the denominator. Thus, the plasma conduction loss is evaluated as

$$\overline{P}_L = \frac{\langle 1.5kTn \rangle}{\tau_E} = \frac{A_L n(0)}{\tau_E} = \frac{A_L n(0)}{C_{op}'}\left(\overline{P}_L \times 10^{-6}\right)^S \tag{A-7}$$

where < > means the volume averaging, and

$$A_L = 1.5e\left\{f_D + f_T + 1/\gamma_i + f_\alpha\right\}T_i(0)/\left(1+\alpha_n+\alpha_T\right)$$

$$C_{op}' = \gamma_{ISS} \times 0.079\,t_{2/3}^{0.4}\,\overline{n}_{19}^{\,0.51}[\times 10^{19}\,m^{-3}]B_o^{0.83}[T]\overline{a}^{2.21}R^{0.65}[m]/\left(2\pi R \pi \overline{a}^2\right)^S$$

The plasma conduction loss $\overline{P}_L$ is finally obtained from Eq. (A-7) as

$$\overline{P}_L = \left[\frac{A_L n(0)}{C_{op}'}\right]^{\frac{1}{1-S}}\left(10^{-6}\right)^{\frac{S}{1-S}} \tag{2-10}$$

The confinement time itself is calculated by

$$\tau_E = \frac{A_L n(0)}{\overline{P}_L} = \frac{\left(C_{op}'\right)^{\frac{1}{1-S}}}{\left[A_L n(0) \times 10^{-6}\right]^{\frac{S}{1-S}}} \tag{A-8}$$

The Sudo density limit [25] is given by

$$n(0)_{\lim}[m^{-3}] = \gamma_{SUDO}\frac{0.25 \times 10^{20}}{\gamma_{pr}}\sqrt{\frac{\left\{P_{NET}[W/m^3] \times 10^{-6} V_o\right\}B_o[T]}{\overline{a}^2 R[m]}} \tag{A-9}$$

Here the profile factor is $\gamma_{pr} = \overline{n}/n(0)$, $V_o = 2\pi R \pi \overline{a}^2$ is the plasma volume.

APPENDIX-2. PELLET PENETRATION DEPTH AND ITS DEPENDENCY ON THE TEMPERATURE AND DENSITY PROFILES

As we do not consider the profile change effect by the pellet penetration, we just mention on the pellet penetration depth as a reference. The pellet penetration depth has been calculated using the neutral gas-shielding (NGS) model [31], which is the most widely adopted ablation model, to estimate the pellet ablation in high temperature plasmas. Despite the fact that it only considers the 1-dimensional electron heat flux to the pellet, it indicates the reasonable agreement with a number of experiments [35][36].

[1] Density Profile without the Edge Density Assumed in this Paper

As the edge density is almost zero in the assumed density profile (Figure 2) in this paper, the pellet penetration depth does not depend on the temperature. This is because the pellet is not evaporated at all in the edge regime. The pellet penetration depths for the three sizes of 12mm, 14 mm, and 16 mm are shown in Figure A-1. In all the cases, the pellets are penetrated to the inside of the half minor radius, which could produce the assumed profile in Figure 2.

Although this is the very special case for the pellet penetration point of view because of no edge density, we have calculated the high-density ignition using this assumed profile for simplicity.

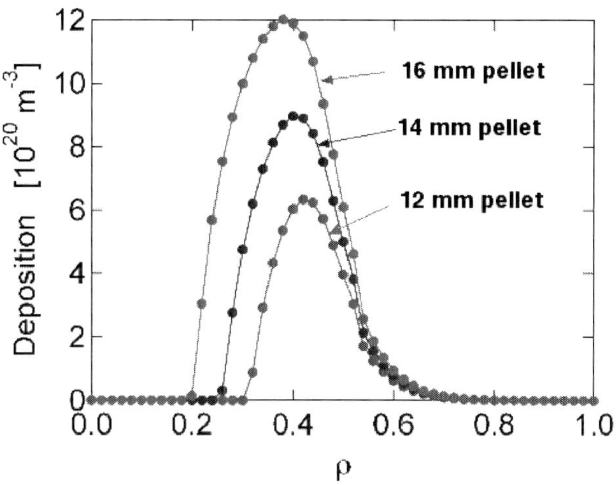

Figure A-1. Deposition profiles by 12, 14, and 16 mm pellet injection with the injection speed of 1000 m/s for the assumed density and temperature profiles (Figure 2) in this study.

[2] Flat Density Profile with Various Temperatures

For further study, dependencies of the pellet penetration depth on the various temperatures are calculated for the flat density profile. We show one typical example here. For the flat density profile with 5×10^{20} m^{-3} and the parabolic temperature profile, a pellet is evaporated at the plasma edge due to the high density and temperature. Therefore, the pellet deposition depth depends on the temperature.

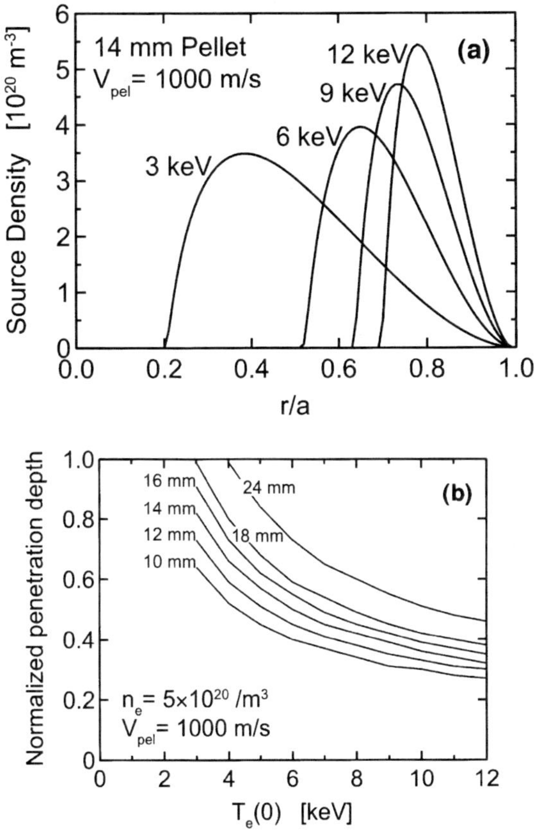

Figure A-2. (a) Deposition profiles of 14 mm pellet for various temperatures, and (b) Dependencies of the normalized penetration depth on the various temperatures for the pellet sizes of 10 ~ 24 mm. The density profile is flat. (Note: Normalized penetration depth of 1.0 means the plasma center).

Figure A-2-(a) shows the deposition profiles of the 14 mm pellet in the central temperature range of 3 keV and 12 keV. Although the penetration depth become shallow as the target plasma temperature rise, it is predicted that the pellet can provide the particles inside the normalized minor radius of ρ= 0.7 with the high temperature of T(0)= 12 keV.

Temperature dependence of the pellet penetration depth, which is normalized by the minor radius, with various pellet size between 10 mm and 24 mm are shown in Figure A-2-(b). The pellet penetration depth can be described approximately by a negative power function of the central temperature and, therefore, change of the penetration depth with temperature becomes moderated in the high temperature regime. Pellet penetration into around ρ = 0.7 is basically possible in high temperature regime, which is envisioned in high-density ignition scenario. For example, in the temperature range of 6~8 keV, the pellet penetration depth is around ρ=0.5 to 0.6. Therefore, it could produce the profile similar to the assumed one, although it may be slightly wider.

Helical devices have an intrinsic advantage to operating with high density because of net current free plasma, and density perturbation due to the pellet injection has less impact on the plasma confinement properties. In addition to this, relatively large pellets can be injected in high-density operations because more particles are retained in the high-density plasma. A helical device with high-density operation has high tolerance to the large size pellet injection. This fact permits a deeper pellet penetration to the plasma.

We note that although a plasmoid drift effect is also observed in helical device [37], the drift effect is not significant compared to tokamak devices because the confinement field is generated by external coils [38], and hence the drift effect is not taken into account in above calculations.

REFERENCES

[1] Ohyabu, N., Morisaki, T., Masuzaki, S., Sakamoto, R., et al., (2006). *Phy. Rev. Lett.*, *97*, 055002-1

[2] Sakamoto, R., et al., (2009). *Nucl. Fusion*, *49*, 1

[3] McCormic, K., Gigull, P., Burhenn, R., et al., (2002). *Phy. Rev. Lett.*, *89*, 015001-1

[4] Mitarai, O., Sagara, A., Ohyabu, N., Sakamoto, R., Komori, A. & Motojima, O. (2007). Plasma and Fusion Research, *Rapid Communication*, *2*, 021-1-3

[5] Mitarai, O., Sagara, A., Ohyabu, N., Sakamoto, R., Komori, A. & Motojima, O. (2009). *Fusion Science and Technology*, *56*, 1495

[6] Mitarai, O., Sagara, A., Ashikawa, N., Sakamoto, R., et al., (2008). "The low temperature and high density ignition in the helical reactor FFHR2m based on LHD experiments", in 22nd IAEA Fusion Energy Conference, (13-8, Oct. 2008, Geneve, Switzerland) FT/P3-19

[7] Mills, R. G. (1969). "The Problems of Control of Thermonuclear Reactors," Proc. Symp. Engineering of Fusion Reactors, LA-4250, Los Alamos National Laboratory, B1-1

[8] Ohta, M., Yamato, H. & Mori, S. (1972). "Thermal instability and control of fusion reactor, 1971 " in Plasma Physics and Controlled Nuclear Fusion Research 1971 (*Proc. 4th Int. Conf. Madison*, 1971), Vol. *3*, p. 423, IAEA, Vienna

[9] Fujisawa, T. (1978). *Nuclear Fusion*, *14*, 173

[10] Bromberg, L., Fisher, J. L. & Cohn, D. R. (1980). *Nuclear Fusion*, *20*, 203.

[11] Borrass, K. (1980). "Burn stabilization of an ignited tokamak in a static or pseudo-static vertical field," Proc. 4th Topl. Mtg. Technology of Controlled Nuclear Fusion. King-of Prussia, Pennsylvania, Oct 14-17, American Nuclear Society, p154

[12] Okamoto, M., Ohnishi, M. & Hirano, K. (1982). *Japanese J. of Applied Physics*, *21*, L201

[13] Kolesnichenko, Ya. I. & Reznik, S. N. (1978). *Nuclear Fusion*, *18*, 1535.

[14] Petrie, T. W. & Rawls, J. M. (1980). *Nuclear Fusion*, *20*, 1461

[15] Maki, K. (1986). *Fusion Technology*, *10*, 70.

[16] Mandrekas, J. & Stacy Jr., W. M. (1991). *Fusion Technology*, *19*, 57

[17] Bebhan, E. & Vieth, U. (1996). *Nucl. Fusion*, *37*, 251

[18] Hui, W., Bamieth, B. & Miley, G. H. (1994). *Fusion Technology*, *25*, 318

[19] Schuster, E., Krstic, M. & Tynan, G. (2003). *Fusion Science and Technology*, *43*, 18.

[20] Vitela, J. E. & Martinell J. J. (1998). *Plasma Physics & Controlled Fusion*, *40*, 295

[21] Mitarai, O., Sagara, A., Sakamoto, R., Yanagi, N., Goto, T., Imagawa, S., Kaneko, O. & Komori, A. (2010). "The high density ignition in FFHR helical reactor by neutral beam injection (NBI) heating" IAEA-FEC-FTP/P6-19 (2010, Daejon)

[22] Mitarai, O., Sagara, A., Sakamoto, R., Yanagi, N. & Goto, T. (2013). *Fusion Engineering and Design*, 88, 1046

[23] Mitarai, O., Sagara, A., Sakamoto, R., Ohyabu, N. A., Komori, K. & Motojima, O. (2010). *Plasma and Fusion Research*, *5*, S1001

[24] Stroh, U., Murakami, M., Dory, R. A., Yamada, H., Okamura, S., Sano F. & Obiki, T. (1996). *Nucl. Fusion*, *36*, 1063

[25] Sudo S., Takeiri Y., Zushi H., et al., (1990). *Nucl. Fusion*, *30*, 11

[26] Miyazawa J., Sakamoto R., Masuzaki S., et al., (2008). *Nucl. Fusion*, *48*, 1-7

[27] Mitarai O., Sagara A., Chikaraishi H., et al., (2007). *Nucl. Fusion*, *47*, 1411-1417

[28] Sagara A., Imagawa S., Mitarai O., Dolan T., et al., (2005). *Nucl Fusion*, *45*, 258

[29] Mitarai, O., Oda, A., Sagara, A., Yamazaki, K. & Motojima, O. (2004). *Fusion Engineering and Design*, *70*, 247-267

[30] Souers, P. C. (1986). "Hydrogen Properties for Fusion Energy", University of California Press p80

[31] Parks, P. B. & Turnbull, R. J. (1978). *Physics of Fluids*, *21*, 1735-1741

[32] Inoue, T., et al., (2006). *Fusion Engineering and Design*, *81*, 1291.

[33] Yanev, R. K., et al., (1989). *Nucl. Fusion*, *29*, 2125

[34] Mitarai, O. & Muraoka, K. (1997). *Nucl. Fusion*, *37*, 1523-1540.

[35] Baylor, L. R. et al., (1997). *Nuclear Fusion*, *37*, 445-450

[36] Hoshino, M. et al., (2006). *Plasma and Fusion Research*, *1*, 033

[37] Sakamoto, R. et al., (2001). *Nuclear Fusion*, *41*, 381-386

[38] Sakamoto, R. et al., (2002). 29th European Physical Society Conference on Plasma Physics and Controlled Fusion, *Europhysics conference abstracts*, *26B*, P1.074 http://epsppd.epfl.ch/Montreux/pdf/P1_074.pdf.

INDEX

D

E

F

V

W

Y

Z